Number 435

Aloys Krieg

Hecke algebras

Memoirs
of the American Mathematical Society

Providence · Rhode Island · USA

September 1990 · Volume 87 · Number 435 (end of volume) · ISSN 0065-9266

Memoirs of the American Mathematical Society
Number 435

Aloys Krieg

Hecke algebras

Published by the
AMERICAN MATHEMATICAL SOCIETY
Providence, Rhode Island, USA

September 1990 · Volume 87 · Number 435 (end of volume)

1980 *Mathematics Subject Classification* (1985 *Revision*).
Primary 10D12, 10D20, 20C05, 20H30.

Library of Congress Cataloging-in-Publication Data

Krieg, Aloys, 1955-
Hecke algebras/Aloys Krieg.
p. cm. – (Memoirs of the American Mathematical Society, ISSN 0065-9266; no. 435)
Includes bibliographical references.
ISBN 0-8218-2497-X
1. Hecke algebras. I. Title. II. Series.
QA3.A57 no. 435
[QA171]
510 s–dc20 90-40201
[512′. 2] CIP

Subscriptions and orders for publications of the American Mathematical Society should be addressed to American Mathematical Society, Box 1571, Annex Station, Providence, RI 02901-1571. *All orders must be accompanied by payment.* Other correspondence should be addressed to Box 6248, Providence, RI 02940-6248.

SUBSCRIPTION INFORMATION. The 1990 subscription begins with Number 419 and consists of six mailings, each containing one or more numbers. Subscription prices for 1990 are $252 list, $202 institutional member. A late charge of 10% of the subscription price will be imposed on orders received from nonmembers after January 1 of the subscription year. Subscribers outside the United States and India must pay a postage surcharge of $25; subscribers in India must pay a postage surcharge of $43. Each number may be ordered separately; *please specify number* when ordering an individual number. For prices and titles of recently released numbers, see the New Publications sections of the NOTICES of the American Mathematical Society.

BACK NUMBER INFORMATION. For back issues see the AMS Catalogue of Publications.

MEMOIRS of the American Mathematical Society (ISSN 0065-9266) is published bimonthly (each volume consisting usually of more than one number) by the American Mathematical Society at 201 Charles Street, Providence, Rhode Island 02904-2213. Second Class postage paid at Providence, Rhode Island 02940-6248. Postmaster: Send address changes to Memoirs of the American Mathematical Society, American Mathematical Society, Box 6248, Providence, RI 02940-6248.

Printed in the United States of America.
The paper used in this book is acid-free and falls within the guidelines established to ensure permanence and durability. ♾

10 9 8 7 6 5 4 3 2 1 95 94 93 92 91 90

Table of Contents

Abstract

This volume gives an introduction to the algebraic theory of Hecke algebras, which can be viewed as generalizations of group algebras. At first a careful look at the product leads to liftings of the basic isomorphism theorems and of anti-homomorphims from the group level to the attached Hecke algebras.

A generalization of Maschke's classical theorem determines the structure of finite dimensional Hecke algebras. Then several examples of Hecke algebras for finite groups are discussed. In particular the cases of minimal and maximal parabolic subgroups of the general linear group over a finite field are investigated.

Chapter IV refers to the roots of Hecke algebras and therefore contains an explicit description of the Hecke algebra associated with the (elliptic) modular group $SL(2;\mathbb{Z})$. Then the result is generalized to the unimodular group of arbitrary degree. Here the number theoretical background material as for instance the Elementary Divisor Theorem or the multiplicativity of elementary divisors is included. Moreover Tamagawa's Rationality Theorem for a particular formal power series over this Hecke algebra is derived.

Finally the structure of the Hecke algebra associated with the Siegel modular group is determined. The commutation relation with the Siegel Φ-operator is described and the surjectivity of the induced homomorphism of the Hecke algebras is demonstrated.

Key Words: Hecke algebra, parabolic subgroup, Borel subgroup, modular group, unimodular group, Rationality Theorem, Siegel modular group, commutation relation with the Siegel Φ-operator.

Introduction

The purpose of these notes is to give an introduction to the algebraic theory of Hecke algebras, where the reader is only supposed to know elementary Algebra and Linear Algebra. We are going to determine the structure of Hecke algebras and deal with several examples explicitly. In particular we will describe the structure of the most important examples of Hecke algebras, which are relevant for the theory of modular forms and the representation theory of finite groups.

Hecke algebras are interesting mathematical objects for several reasons. In Number Theory or more precisely in the theory of modular forms one studies the linear representation of the particular Hecke algebra in the space of modular forms. These operators reveal multiplicative properties of the Fourier coefficients. The Fourier coefficients of modular forms are always interesting arithmetical functions, as for instance divisor sums, representation numbers of quadratic forms, or the Ramanujan τ-function.

On the other hand Hecke algebras play a decisive role in the representation theory of finite groups. In particular they arise in the decomposition problem of induced characters of finite groups of Lie type.

1. The Development

In 1937 E. Hecke [20] introduced certain endomorphisms of the space of entire elliptic modular forms, which are nowadays called Hecke operators. In the particular case of the discriminant Δ these operators had already been studied by L.J. Mordell [48], but E. Hecke was the first, who recognized the universal meaning of this construction. H. Petersson [52] showed that there exists a basis of the space of modular forms, which consists of simultaneous eigenvectors under all Hecke operators. These are the fundamentals for a lot of applications in Number Theory.

E. Hecke's ideas were generalized to other classes of modular forms. M. Sugawara [59], [60] and H. Maaß [43] developed the first notions of Hecke operators for Siegel modular forms. O. Herrmann [21] investigated Hecke operators for Hilbert modular forms.

About 30 years ago G. Shimura [56] introduced the notion of an abstract Hecke algebra. This is the algebraic setting for all kinds of Hecke operators, which had been studied so far. Hecke operators arise by the linear representation of a particular Hecke algebra in the attached space of modular forms.

In order to define an abstract Hecke algebra, one starts with a group G and a subgroup S, where each double coset relative to S is supposed to decompose into finitely many right cosets relative to S. The product of two double cosets is defined as an integral linear combination of all the double cosets, which are contained in the products of the elements of these double cosets. The coefficients are positive integers arising from the right coset decompositions. Thus the free $\mathbb{Z}$-module with all the double cosets relative to S in G as a basis turns out to be an associative $\mathbb{Z}$-algebra with S as its identity element. It is called the Hecke algebra associated with the pair (S,G). If S is a normal subgroup of G, then the Hecke algebra coincides with the group algebra of the factor group G/S over $\mathbb{Z}$. Of course the ground ring $\mathbb{Z}$ can be replaced by an arbitrary commutative unitary ring.

T. Tamagawa [62] investigated the Hecke algebra associated with the unimodular group of degree n, where the case $n = 2$ refers to elliptic modular forms. G. Shimura [57] determined the structure of the Hecke algebra associated with the Siegel modular group. In both cases the Hecke algebras are commutative. They are polynomial rings in infinitely many indeterminates and therefore do not contain any zero-divisors. I. Satake [53] characterized these Hecke algebras as subalgebras of polynomial algebras.

In the representation theory the Hecke algebra associated with (B,G), where G is a finite group with a (B,N)-pair, was described by N. Iwahori [24] and H. Matsumoto [46]. Over $\mathbb{C}$ this Hecke algebra is isomorphic to the group algebra of the attached Weyl group W. For an arbitrary parabolic subgroup in place of B the corresponding isomorphism is due to C.W. Curtis, N. Iwahori, and R. Kilmoyer [15]. D. Kazhdan and G. Lusztig [27] investigated the meaning of Hecke algebras for representations of Coxeter groups. When dealing with groups over local fields the reader is referred to the work of N. Iwahori and H. Matsumoto [25] as well as D. Kazhdan and G. Lusztig [28].

2. The Contents

In chapter I we introduce the abstract Hecke algebra of a pair (S,G), where S is a subgroup of a group G such that each double coset relative to S in G decomposes into finitely many right cosets relative to S. We give several descriptions of the multiplicities arising in the product of two double cosets. It is possible to lift homomorphisms and anti-homomorphisms from the group level to the algebra level. This leads to the basic isomorphism theorems for groups in the context of Hecke algebras. As an application we obtain a simple sufficient condition for Hecke algebras to be commutative. Finally we introduce a canonical bilinear form. The double cosets are an orthogonal basis with respect to this bilinear form. We describe the exact conditions such that the bilinear form is non-degnerate provided that the ground ring is a field.

In chapter II we describe the structure theory of Hecke algebras for finite groups more precisely. A generalization of Maschke's classical theorem for group algebras says that the Hecke algebra associated with a pair (S,G) over a field F is semi-simple, whenever the characteristic of F does not divide the order of the finite group G. A closer look at Hecke algebras of small dimension yields that every Hecke algebra of dimension ≤ 5 is commutative. In analogy with W. Scharlau's results for group algebras [54] we can study the Hecke algebra endowed with the canonical involution, which is induced by the involution $g \mapsto g^{-1}$ of G.

In chapter III we discuss several examples of Hecke algebras for finite groups explicitly. In particular we describe the structure of the Hecke algebras associated with (S,G), where G is the general linear group over a finite field and S is a minimal resp. a maximal parabolic subgroup. If S is an arbitrary parabolic subgroup, one can apply the results of C.W. Curtis, N. Iwahori, and R. Kilmoyer [15] and show that the Hecke algebra associated with (S,G) is commutative if and only if S is a maximal parabolic subgroup.

Chapter IV refers to E. Hecke's classical situation, i.e., we consider the Hecke pair $(SL(2;\mathbb{Z}),GL^{+}(2;\mathbb{Q}))$. In this case we can describe the products explicitly. We give a few applications to congruence subgroups.

The Hecke algebra associated with $(GL(n;\mathbb{Z}),GL(n;\mathbb{Q}))$ is investigated in chapter V. We start with the number theoretical background material, which is necessary for the investigation, e.g., Elementary Divisor Theorem, multiplicativity of elementary divisors. Following T. Tamagawa [62] we can derive a Rationality Theorem for a formal power series over this Hecke algebra.

Finally we deal with the Hecke algebra related with the Siegel modular group in chapter VI. Moreover we investigate the homomorphism of the Hecke algebras, which arises from the commutation relation with the Siegel Φ-operator. We calculate the images of a set of generators explicitly and thus derive the surjectivity of the homomorphism.

These notes do not contain the applications to modular forms, i.e., Hecke operators, since this would be at least one other book. But all the chapters include guides to the descriptions of the applications in the literature.

3. The Structure

This book has the character of lecture notes. It is intended that one can read the more number theoretical chapters without the chapters on Hecke algebras of finite groups. Hence it is possible to understand the chapters IV, V, and VI without the knowledge of the chapters II and III. It is not even necessary to go through chapter IV before chapters V and VI.

It is hardly possible always to name the person the theorem is due to. Therefore Remarks are included at the end of each chapter. Here the reader can find references to the standard literature as well as to original papers.

The results of each section are numbered consecutively with the number of the section in the first place. The chapter is added in Roman numerals only if a different chapter is cited.

4. Notations

Let $\mathbb{N}$ denote the set of positive integers, $\mathbb{N}_0$ the set of non-negative integers and $\mathbb{Z}$ the set of rational integers. The fields of rational resp. real resp. complex numbers are abbreviated by $\mathbb{Q}$ resp. $\mathbb{R}$ resp. $\mathbb{C}$. Given an arbitrary ring R and $m,n \in \mathbb{N}$ let Mat(m,n;R) denote the set of all matrices having m rows and n columns and entries in R. We use the abbreviations Mat(n;R) := Mat(n,n;R) for the ring of square matrices and R^m := Mat(m,1;R) for the space of column vectors. On the other hand $A^{(m,n)}$ also stands for an $m \times n$ matrix as well as $A^{(n)}$ for $A^{(n,n)}$. If $A_1,\ldots,A_n$ are square matrices $\operatorname{diag}(A_1,\ldots,A_n)$ is defined to be that square matrix with $A_1,\ldots,A_n$ as diagonal blocks and 0 's otherwise. The letter $I = I^{(n)}$ is reserved for the $n \times n$ identity matrix diag(1,1,...,1), whereas 0 stands for a rectangular matrix of suitable size having only 0 's as entries. If the ring R has an identity element, we denote the group of units in the ring Mat(n;R) by GL(n;R). If A is a rectangular matrix, then A' stands for its transpose. Sym(n;R) is defined to be the set of all symmetric $n \times n$ matrices over R. Given $a,b \in R$ then $a \mid b$ means that a is a divisor of b. Moreover we use the abbreviations g.c.d. for the greatest common divisor and l.c.m. for the least common multiple. The identical map on a set S is denoted by id. By S_n we mean the symmetric group of n letters, i.e., the group of bijective maps of $\{1,2,\ldots,n\}$ onto itself. The Kronecker symbol is denoted by δ_{jk}. Finally $\#(S)$ is defined to be the number of elements of the set S.

5. Acknowledgements

This book is based on courses I gave at the Panjab University, Chandigarh (India) in 1987 and at the University of California, San Diego (USA) in 1988. My thanks are due to Professor A. Terras (San Diego), who encouraged me to write these notes. Moreover I thank Professor M. Koecher (Münster) for his helpful suggestions.

Chapter I

The Abstract Hecke Algebra

The purpose of this chapter is to derive basic properties of abstract Hecke algebras. Hecke algebras can be viewed as generalizations of group algebras, so we start with group algebras. The Hecke condition is the necessary condition for the construction of Hecke algebras. Then we give several descriptions of the product from various points of view.

The basic isomorphism theorems for groups can be extended to isomorphism theorems for Hecke algebras. We can also lift anti-homomorphisms from the group level to the algebra level. As an application we obtain a sufficient criterion for the Hecke algebra to be commutative. This result will be applied in all the examples.

Finally we deal with a symmetric bilinear form, which will be applied in the next chapter in order to derive a structure theorem. The double cosets form an orthogonal basis with respect to this bilinear form. We derive the exact conditions on the ground field such that the bilinear form is non-degenerate.

1. The Group Algebra

Throughout the whole chapter let R denote a commutative ring with identity element $1 = 1_R$ and G a multiplicatively written group with identity element e. We set

$$R[G] = \{\varphi : G \longrightarrow R \mid \varphi(g) \neq 0 \text{ only for finitely many } g \in G\}.$$

The elements $\varphi \in R[G]$ have finite *support*

$$\operatorname{supp}(\varphi) := \{g \in G \mid \varphi(g) \neq 0\}.$$

Given $\alpha \in R$ and $\varphi, \psi \in R[G]$ we define mappings $G \longrightarrow R$ via

$$(\alpha\varphi)(g) := \alpha\varphi(g),$$

$$(\varphi+\psi)(g) := \varphi(g) + \psi(g),$$

(1) $$(\varphi \cdot \psi)(g) := \sum_{h \in G} \varphi(h)\psi(h^{-1}g) = \sum_{h_1, h_2 \in G, h_1 h_2 = g} \varphi(h_1)\psi(h_2) .$$

Received by the editors March 31, 1989.

Since φ and ψ have finite support, the sums in (1) are finite. $\varphi\cdot\psi$ belongs to $R[G]$ again in view of

$$\# \operatorname{supp}(\varphi\cdot\psi) \leq \# \operatorname{supp}(\varphi) \; \# \operatorname{supp}(\psi) .$$

Given $g \in G$ we define the Kronecker-delta $\delta_g \in R[G]$ by

$$\delta_g(h) := \begin{cases} 1 & \text{if } h = g , \\ 0 & \text{if } h \neq g . \end{cases}$$

(1.1) Theorem. *Let* R *be a commutative unitary ring and* G *a group with identity element* e . *Then* $R[G]$ *is an associative* R*-algebra with* δ_e *as its identity element. The mappings* δ_g , $g \in G$, *form a basis of the* R*-module* $R[G]$ *and satisfy*

$$\delta_g \cdot \delta_h = \delta_{gh} \quad \textit{for } g,h \in G .$$

The algebra $R[G]$ *is commutative if and only if* G *is abelian.*

Definition. $R[G]$ is called the *group algebra of* G *over* R .

The results are well-known and contained in most books on Algebra, cf. N. Jacobson [26] , p.123, p.388, or T. Hungerford [23] .

Identifying δ_g with g we can also say that $R[G]$ consists of all formal finite linear combinations of the form

$$\sum_{g\in G} \varphi(g)g , \quad \varphi(g) \in R , \ \varphi(g) \neq 0 \ \text{ only for finitely many } g \in G .$$

The product can then be given in the form

$$\text{(2)} \qquad \Big(\sum_{g\in G} \varphi(g)g\Big) \cdot \Big(\sum_{h\in G} \psi(h)h\Big) = \sum_{g,h\in G} \varphi(g)\psi(h)gh.$$

It is not necessary to start with a group. Taking the product (2) it suffices to consider a semi-group S in place of G . In this case $R[S]$ turns out to be an R-algebra again and is called the *semi-group algebra of* S *over* R . Clearly $R[S]$ possesses an identity element if and only if S is a monoid.

It is not difficult to determine the center of a group algebra. More generally the *center* of an algebra $\mathcal{A}$ is defined by

$$\operatorname{cent} \mathcal{A} := \{ a \in \mathcal{A} \mid ab = ba \ \text{ for all } \ b \in \mathcal{A} \}$$

and proves to be a subalgebra of $\mathcal{A}$. The description of $\operatorname{cent} R[G]$ is clearly related with the data of the group G . Given $g \in G$ we define the *conjugacy class* of g in G by

$$\{ hgh^{-1} \mid h \in G \} .$$

Two conjugacy classes are disjoint or equal.

(1.2) Lemma. *Let* R *be a commutative unitary ring and* G *a group. Let* $\mathcal{C}(G)$ *denote the collection of all finite conjugacy classes in* G . *Then the elements*

$$T_C := \sum_{g \in C} g \quad , \quad C \in \mathcal{C}(G) ,$$

form an R-*basis of the center of the group algebra* R[G] .

Proof. Given $h \in G$ the definition of a conjugacy class leads to

$$hCh^{-1} = C \quad , \quad C \in \mathcal{C}(G) .$$

Hence $T_C \in \operatorname{cent} R[G]$, $C \in \mathcal{C}(G)$, follows. According to $\operatorname{supp} T_C = C$ and the fact that different conjugacy classes are disjoint, the elements T_C , $C \in \mathcal{C}(G)$, are linearly independent over R . Let

$$T = \sum_{g \in G} t(g)g \in \operatorname{cent} R[G] .$$

Then $h \cdot T = T \cdot h$ implies

$$t(hgh^{-1}) = t(g) \quad \text{for all } g,h \in G .$$

Hence the coefficients of the elements of a conjugacy class coincide. Since T has finite support, T becomes a finite linear combination of the T_C , $C \in \mathcal{C}(G)$. □

For the general theory of group algebras the reader is referred to D.S. Passman [51] .

(1.3) Exercises. Let F be a field, whenever it occurs.

a) Let G be an infinite cyclic group. Show that F[G] is a principal ideal domain.

b) Show that the map

$$T = \sum_{g \in G} t(g)g \mapsto \hat{T} = \sum_{g \in G} t(g)g^{-1}$$

is an anti-automorphism of R[G] , i.e., it is an automorphism of the R-module and satisfies

$$\widehat{(T_1 \cdot T_2)} = \hat{T}_2 \cdot \hat{T}_1 \quad \text{for } T_1, T_2 \in R[G] .$$

c) Let G be finite. Show that every element of $\mathbb{Z}[G]$ is integral over $\mathbb{Z}$, i.e., root of a monic polynomial $f(X) \in \mathbb{Z}[X]$ (cf. [42] , chap. IX). Conclude that the elements of F[G] are algebraic over F for an arbitrary field F .

d) Let S be a subgroup of G and define

$$\pi_S : R[G] \to R[S] ,$$
$$\sum_{g \in G} t(g)g \mapsto \sum_{g \in S} t(g)g .$$

Given $T_1, T_2 \in R[G]$, $U \in R[S]$, and $\alpha, \beta \in R$ show that

(i) $\pi_S(\alpha T_1 + \beta T_2) = \alpha\,\pi_S(T_1) + \beta\,\pi_S(T_2)$.

(ii) $\pi_S(U \cdot T_1) = U \cdot \pi_S(T_1)$, $\pi_S(T_1 \cdot U) = \pi_S(T_1) \cdot U$.

(iii) π_S is an algebra homomorphism if and only if $S = G$.

e) Determine the center of the group algebra $R[S_3]$, where S_3 is the symmetric group of 3 letters.

2. Formal Linear Combinations of Right and Double Cosets

Now we slightly generalize the formal procedure described in section 1 . Therefore let a group G and a subgroup S of G be given. For $g \in G$ the *right coset of* g *relative to* S is defined by

$$Sg = \{sg \mid s \in S\},$$

the *left coset of* g *relative to* S is given by

$$gS = \{gs \mid s \in S\},$$

and the *double coset of* g *relative to* S is

$$SgS = \{s_1 g s_2 \mid s_1, s_2 \in S\}.$$

Note that the right and left cosets possess as many elements as S does. But the orders of double cosets do not coincide in general. An explicit example will be given in section 5 . But one can also see this more generally. One has

$$SgS = S \quad \text{for} \quad g \in S .$$

If S is finite, we get for $g \in G$

$$\#(SgS) = \#(S) \iff Sg = gS .$$

Recall that for $g,h \in G$ the following conditions are equivalent:

(i) $Sg = Sh$.

(ii) $Sg \cap Sh \neq \emptyset$.

(iii) There exists $s \in S$ satisfying $sg = h$.

The analogue is of course true for left cosets. A straightforward verification also yields the equivalence of

(i) $SgS = ShS$.

(ii) $SgS \cap ShS \neq \emptyset$.

(iii) There exist $s_1, s_2 \in S$ satisfying $s_1 g s_2 = h$.

Thus we can define the following equivalence relations on G :

$$g \underset{r}{\sim} h \iff Sg = Sh ,$$
$$g \underset{l}{\sim} h \iff gS = hS ,$$
$$g \underset{d}{\sim} h \iff SgS = ShS .$$

Let X be a subset of G , which consists of full eqivalence classes in each case, e.g., $X = G$. Then "g: $S\backslash X$" means that g runs through a set of representatives of the right cosets relative to S in X , i.e., of $X/\underset{r}{\sim}$. Similarly "g : X/S" denotes a set of representatives of the left cosets relative to S in X , i.e., of $X/\underset{l}{\sim}$. Finally "g : $S\backslash X/S$" stands for a set of representatives of the double cosets relative to S in X , i.e., of the equivalence classes $X/\underset{d}{\sim}$.

Let $\mathcal{R}(S,G)$ or more precisely $\mathcal{R}_R(S,G)$ denote the free R-module with all the right cosets Sg relative to S in G as a basis. Thus $\mathcal{R}(S,G)$ consists of the elements

(1) $$U = \sum_{g:S\backslash G} u(Sg)Sg ,$$

where $u(Sg) \in R$ and $u(Sg) \neq 0$ only for finitely many right cosets Sg . Given

$$V = \sum_{g:S\backslash G} v(Sg)Sg \in \mathcal{R}(S,G)$$

and $\alpha \in R$, the definition of a free R-module means

$$U = V \iff u(Sg) = v(Sg) \text{ for all } g \in G ,$$
$$U + V = \sum_{g:S\backslash G} [u(Sg) + v(Sg)]\, Sg \in \mathcal{R}(S,G) ,$$
$$\alpha U = \sum_{g:S\backslash G} [\alpha u(Sg)]\, Sg \in \mathcal{R}(S,G) .$$

Moreover define a linear form on $\mathcal{R}(S,G)$ via

$$\operatorname{ind} U := \sum_{g:S\backslash G} u(Sg) \in R .$$

G acts on $\mathcal{R}(S,G)$ by multiplication from the right hand side. Given $h \in G$ one gets $r_h \in \operatorname{End} \mathcal{R}(S,G)$ given by

$$r_h(U) := Uh = \sum_{g:S\backslash G} u(Sg)Sgh .$$

Thus we obtain an R-submodule of $\mathcal{R}(S,G)$ by

$$\mathcal{R}(S,G)^S = \{U \in \mathcal{R}(S,G) \mid r_s(U) = U \text{ for all } s \in S\}$$

An element U of the form (1) belongs to $\mathcal{R}(S,G)^S$ if and only if

$$u(SgS) = u(Sg) \quad \text{for all} \quad g \in G\ ,\ s \in S\ .$$

In the same way we define the free R-module $\mathcal{H}(S,G)$ or more precisely $\mathcal{H}_R(S,G)$, consisting of all formal finite linear combinations of the double cosets SgS relative to S in G. The elements of $\mathcal{H}(S,G)$ therefore have the form

$$T = \sum_{g:S\backslash G/S} t(SgS)\, SgS\ ,$$

where $t(SgS) \in R$ and $t(SgS) \neq 0$ only for finitely many double cosets SgS in G. Given $\alpha \in R$ and

$$\bar{T} = \sum_{g:S\backslash G/S} \bar{t}(SgS)\, SgS \in \mathcal{H}(S,G)\ ,$$

the definition means

$$T = \bar{T} \iff t(SgS) = \bar{t}(SgS) \quad \text{for all} \quad g \in G\ ,$$

$$T + \bar{T} = \sum_{g:S\backslash G/S} [t(SgS) + \bar{t}(SgS)]\, SgS \in \mathcal{H}(S,G)\ ,$$

$$\alpha T = \sum_{g:S\backslash G/S} [\alpha\, t(SgS)]\, SgS \in \mathcal{H}(S,G)\ .$$

Now it turns out to be essential to establish connections between these R-modules. In the next section we will derive a condition on S and G, which allows us to identify $\mathcal{R}(S,G)^S$ with $\mathcal{H}(S,G)$.

In order to see the connection with group algebras suppose that S is normal in G, i.e., $Sg = gS = SgS$ for all $g \in G$. In this case we have

$$\mathcal{R}(S,G) = \mathcal{R}(S,G)^S = \mathcal{H}(S,G) = R[G/S]\ .$$

So far we have not given any motivation. Why do we study all these formal linear combinations? We explain what is typical for all the applications we have in mind, e.g., Hecke operators for all kinds of automorphic forms. Consider a domain D and for the sake of simplicity a certain complex vector space V of functions $\varphi : D \to \mathbb{C}$. Suppose that G acts on V, i.e., given $g, h \in G$, $\varphi, \psi \in V$, $\alpha, \beta \in \mathbb{C}$, then a function $\varphi \mid g$ is defined in V and one has

$$(\alpha\varphi + \beta\psi) \mid g = \alpha(\varphi \mid g) + \beta(\psi \mid g)\ , \quad (\varphi \mid g) \mid h = \varphi \mid (gh)\ .$$

One is interested in the subspace

$$V^S := \left\{ \varphi \in V \mid \varphi \mid s = \varphi \quad \text{for all} \quad s \in S \right\}.$$

Set $R = \mathbb{C}$ and consider $U \in \mathcal{R}(S,G)$ of the form (1) and $\varphi \in V^S$. Then

$$\text{(2)} \qquad \varphi \mid U := \sum_{g:S\backslash G} u(Sg)\, \varphi \mid g \in V$$

is well-defined. Now one easily checks $\varphi \mid U \in V^S$, whenever $U \in \mathcal{R}(S,G)^S$. Hence (2) induces a linear mapping

(3) $$\mathcal{R}(S,G)^S \to \operatorname{End} V^S ,$$

where $\operatorname{End} V^S$ denotes the set of all endomorphisms of the vector space V^S. Hence the main purpose is to construct endomorphisms of V^S from the elements in G.

(2.1) Exercises. Let S_1, S_2 be two subgroups of a group G. Consider the (S_1,S_2)-double cosets

$$S_1gS_2 := \{ s_1gs_2 \mid s_1 \in S_1 , \ s_2 \in S_2 \} , \ g \in G .$$

a) Characterize the equivalence relation given by (S_1,S_2)-double cosets.
b) Find an example of G, S_1, S_2 with $S_1 \neq S_2$ and

$$S_1gS_2 = S_2gS_1 \quad \text{for some } g \in G .$$

c) Show that the map $G \to G$, $g \mapsto g^{-1}$, induces a bijection between the collections of all (S_1,S_2)-double cosets and all (S_2,S_1)-double cosets in G.
d) Let S be a finite subgroup of a group G. Show that the normalizer $N(S)$ of S in G is equal to

$$N(S) := \{ g \in G \mid gS = Sg \} = \{ g \in G \mid \#(S\backslash SgS) = 1 \} .$$

3. Hecke Pairs

The Hecke condition says that every double coset decomposes into finitely many right cosets. The Hecke condition will allow us to identify the R-modules $R(S,G)^S$ and $\mathcal{H}(S,G)$. At first we describe the meaning of this condition.

(3.1) Lemma. *Let* S *be a subgroup of a group* G *and* $g \in G$. *Then one has*

(i) $$SgS = \bigcup_{s : (S \cap g^{-1}Sg)\backslash S} Sgs = \bigcup_{s : S/(S \cap gSg^{-1})} sgS$$

as disjoint unions.

(ii) $$\#(S\backslash SgS) = [S : S \cap g^{-1}Sg] , \quad \#(SgS/S) = [S : S \cap gSg^{-1}] .$$

Proof. (i) Set $\bar{S} = S \cap g^{-1}Sg$. So we show that

(1) $$\bigcup_{s : \bar{S}\backslash S} Sgs = SgS$$

holds as a disjoint union. In view of $g\bar{s}g^{-1} \in S$ for $\bar{s} \in \bar{S}$ the left hand side of (1) does not depend on the choice of the representatives and is clearly contained in SgS. If

$\bar{g} = s_1 g s_2$ with $s_1, s_2 \in S$, then s_2 belongs to a right coset $\bar{S}s$, i.e., $s_2 = \bar{s}s$ with some $\bar{s} \in \bar{S}$. Thus

$$\bar{g} = s_1(g\bar{s}g^{-1})gs \in Sgs$$

follows. Hence equality holds in (1).

Now let $s_1, s_2 \in S$ such that $Sgs_1 = Sgs_2$. This leads to $sgs_1 = gs_2$ for some $s \in S$, whence $g^{-1}sg = s_2 s_1^{-1} \in \bar{S}$. Thus we get $\bar{S}s_1 = \bar{S}s_2$ and (1) holds as a disjoint union.

Apply analogous arguments to the second part.

(ii) Use (i). □

If S is finite then (3.1) gives a formula for the number of elements in a double coset, namely

$$\begin{aligned} \#(SgS) &= \#(S)\,[S : S \cap g^{-1}Sg] = \#(S)\,\#(S\backslash SgS) \\ &= \#(S)\,[S : S \cap gSg^{-1}] = \#(S)\,\#(SgS/S). \end{aligned}$$

But in general the numbers of left and right cosets in a double coset do not coincide, whenever S in infinite (cf. (3.5f)).

Two subgroups S_1 and S_2 of a group G are called *commensurable* if the intersection $S_1 \cap S_2$ has finite index in both S_1 and S_2. In this case we write $S_1 \sim S_2$. We show that commensurability is an equivalence relation on the set of all subgroups of a given group G.

(3.2) Lemma. *Let* S_1, S_2, S_3 *be subgroups of a group* G. *If* $S_1 \sim S_2$ *and* $S_2 \sim S_3$ *then* $S_1 \sim S_3$.

Proof. One has

$$[S_1 \cap S_2 : S_1 \cap S_2 \cap S_3] \le [S_2 : S_2 \cap S_3] < \infty,$$

$$\begin{aligned} [S_1 : S_1 \cap S_3] &\le [S_1 : S_1 \cap S_2 \cap S_3] \\ &= [S_1 : S_1 \cap S_2]\,[S_1 \cap S_2 : S_1 \cap S_2 \cap S_3] < \infty. \end{aligned}$$

Analogous arguments yield $[S_3 : S_1 \cap S_3] < \infty$. □

Given a subgroup S of G we define the *commensurator of* S *in* G by

$$\mathrm{Com}(S,G) = \{g \in G \mid g^{-1}Sg \sim S\}.$$

The structure of the set $\mathrm{Com}(S,G)$ is described in

(3.3) Proposition. *Assume that* S *is a subgroup of a group* G. *Then the commensurator* $\mathrm{Com}(S,G)$ *is a subgroup of* G *containing* S.

Proof. Clearly $S \subset \mathrm{Com}(S,G)$ holds. If S_1, S_2 are subgroups of G and $g \in G$ one has

(2) $$S_1 \sim S_2 \Rightarrow g^{-1}S_1g \sim g^{-1}S_2g .$$

Given $g \in \mathrm{Com}(S,G)$ then $S \sim g^{-1}Sg$ implies $gSg^{-1} \sim S$ by virtue of (2), hence we get $g^{-1} \in \mathrm{Com}(S,G)$. If $g_1, g_2 \in \mathrm{Com}(S,G)$ then $g_1^{-1}Sg_1 \sim S$ leads to $g_2^{-1}(g_1^{-1}Sg_1)g_2 \sim g_2^{-1}Sg_2 \sim S$. But (3.2) yields $g_1g_2 \in \mathrm{Com}(S,G)$. □

Note that S and $\mathrm{Com}(S,G)$ fail to be commensurable in general. In chapter IV we will for instance deal with $S = SL(2;\mathbb{Z})$ and $G = SL(2;\mathbb{Q})$. Then $G = \mathrm{Com}(S,G)$ will be proved, but S does not have finite index in G.

Given $g \in \mathrm{Com}(S,G)$ then $\#(S\backslash SgS) = \#(Sg^{-1}S/S)$ holds. Thus (3.1) and (3.3) lead to

(3.4) Corollary. *Let* S *be a subgroup of* G.

a) $$\mathrm{Com}(S,G) = \{g \in G \mid \#(S\backslash SgS) < \infty \text{ and } \#(SgS/S) < \infty\}.$$

b) *If* $\#(S\backslash SgS) < \infty$ *holds for all* $g \in G$, *then one has*

$$G = \mathrm{Com}(S,G).$$

c) *If* $\#(SgS/S) < \infty$ *holds for all* $g \in G$, *then one has*

$$G = \mathrm{Com}(S,G).$$

Definition. Let S be a subgroup of a group G. We say that (S,G) satisfies the *Hecke condition* and call (S,G) a *Hecke pair*, if

$$G = \mathrm{Com}(S,G).$$

(3.5) Examples. a) If S is a finite subgroup of a group G, then (S,G) is obviously a Hecke pair.

b) If S is a normal subgroup of a group G, then (S,G) is a Hecke pair.

c) If (S,G) is a Hecke pair and $\bar{S}$ is a subgroup of G, which is commensurable with S, then $(\bar{S},G)$ is a Hecke pair, too. Using (2), we conclude for all $g \in G$

$$\bar{S} \sim S \sim g^{-1}Sg \sim g^{-1}\bar{S}g .$$

d) In chapter V we will show that $\big(GL(n;\mathbb{Z}), GL(n;\mathbb{Q})\big)$ is a Hecke pair, where

$$GL(n;\mathbb{Z}) = \{g \in \mathrm{Mat}(n;\mathbb{Z}) \mid \det g = \pm 1\}.$$

e) $\big(GL(2;\mathbb{Z}), GL(2;\mathbb{R})\big)$ is not a Hecke pair. In order to see this, note that the matrices $\begin{pmatrix}1 & n\\ 0 & \pi\end{pmatrix}$, $n \in \mathbb{Z}$, belong to mutually distinct right cosets but to a single double coset in view of

$$\begin{pmatrix}1 & n\\ 0 & \pi\end{pmatrix} = \begin{pmatrix}1 & 0\\ 0 & \pi\end{pmatrix}\begin{pmatrix}1 & n\\ 0 & 1\end{pmatrix},$$

$$\begin{pmatrix}1 & m\\ 0 & \pi\end{pmatrix}\begin{pmatrix}1 & n\\ 0 & \pi\end{pmatrix}^{-1} = \begin{pmatrix}1 & (m-n)\pi^{-1}\\ 0 & 1\end{pmatrix} \in GL(2;\mathbb{Z}) \iff m = n .$$

f) Consider the subgroup

$$S = \left\{\begin{pmatrix}1 & n\\ 0 & 1\end{pmatrix} \mid n \in \mathbb{Z}\right\}$$

of GL(2;Q) . A simple calculation yields

$$G = \mathrm{Com}\big(S, GL(2;\mathbb{Q})\big) = \left\{\begin{pmatrix}a & b\\ 0 & d\end{pmatrix} \mid a,b,d \in \mathbb{Q}\ ,\ ad \neq 0\right\} .$$

Now let $a,b,d \in \mathbb{Z}$ with $ad \neq 0$. One easily gets

$$S\begin{pmatrix}a & b\\ 0 & d\end{pmatrix}S = \left\{\begin{pmatrix}a & b+nd+ma\\ 0 & d\end{pmatrix} \mid n,m \in \mathbb{Z}\right\} .$$

This leads to

$$\#\left(S\backslash S\begin{pmatrix}a & b\\ 0 & d\end{pmatrix}S\right) = \left|\frac{d}{\text{g.c.d.}(a,d)}\right| ,$$

$$\#\left(S\begin{pmatrix}a & b\\ 0 & d\end{pmatrix}S/S\right) = \left|\frac{a}{\text{g.c.d.}(a,d)}\right| .$$

In (3.5f) we dealt with an example, where the numbers of left and right cosets in a double coset do not coincide. But the relation between these numbers is not "irregular" .

(3.6) Proposition. *Let* (S,G) *be a Hecke pair. Then the map*

$$\lambda : G \to \mathbb{Q}^+ ,\ \lambda(g) := \frac{\#(S\backslash SgS)}{\#(SgS/S)} ,$$

is a homomorphism of the groups with

$$S \subset \text{kernel } \lambda .$$

Proof. Let $\mathcal{C}(S)$ denote the set of all subgroups of G , which are commensurable with S . Given $S_1,S_2 \in \mathcal{C}(S)$ we define

$$\lambda(S_1/S_2) := [S_1 : \bar{S}][S_2 : \bar{S}]^{-1} ,$$

where $\bar{S} \in \mathcal{C}(S)$, $\bar{S} \subset S_1 \cap S_2$. Clearly the definition does not depend on the choice of $\bar{S}$, one can for instance take $\bar{S} = S_1 \cap S_2$. Due to (3.2) it is easy to see that

$$\lambda(S_1/S_2)\lambda(S_2/S_3) = \lambda(S_1/S_3) , \qquad \lambda(g^{-1}S_1g/g^{-1}S_2g) = \lambda(S_1/S_2)$$

for all $S_1,S_2,S_3 \in \mathcal{C}(S)$, $g \in G$. Given $g \in G$ one uses (3.1) in order to verify

$$\lambda(g) = \frac{[S : S \cap g^{-1}Sg]}{[S : S \cap gSg^{-1}]} = \frac{[S : S \cap g^{-1}Sg]}{[g^{-1}Sg : g^{-1}Sg \cap S]} = \lambda(S/g^{-1}Sg) .$$

This leads to

$$\begin{aligned}\lambda(g_1g_2) &= \lambda(S/g_2^{-1}g_1^{-1}Sg_1g_2) = \lambda(S/g_2^{-1}Sg_2)\lambda(g_2^{-1}Sg_2/g_2^{-1}g_1^{-1}Sg_1g_2)\\ &= \lambda(S/g_2^{-1}Sg_2)\lambda(S/g_1^{-1}Sg_1) = \lambda(g_2)\lambda(g_1)\end{aligned}$$

for $g_1,g_2 \in G$. □

In the example (3.5f) we obtain

$$\lambda\left(\begin{pmatrix} a & b \\ 0 & d \end{pmatrix}\right) = \left| \frac{d}{a} \right| \quad \text{for } a,b,d \in \mathbb{Q} \text{ , } ad \neq 0 \text{ .}$$

Now assume that (S,G) is a Hecke pair. Given

$$T = \sum_{g:S\backslash G/S} t(SgS)\, SgS \in \mathcal{H}(S,G) \text{ ,}$$

we define

$$\iota(T) := \sum_{g:S\backslash G} t(SgS)\, Sg \in \mathcal{R}(S,G) \text{ .}$$

Note that the definition is obviously independent of the choice of the representative in each right coset. Since (S,G) is a Hecke pair, it is clear that $\iota(T)$ becomes a finite linear combination and thus belongs to $\mathcal{R}(S,G)$. In particular $\iota(SgS)$ equals the formal sum of all the right cosets relative to S , which are contained in the double coset SgS .

In this case we define

$$\text{(3)} \qquad \text{ind } T := \text{ind } \iota(T) = \sum_{g:S\backslash G} t(SgS) = \sum_{g:\, S\backslash G/S} t(SgS)\, \#(S\backslash SgS) \in R \text{ .}$$

In the particular case of a double coset, ind SgS equals the number of right cosets in SgS considered as an element of the ground ring R , i.e.,

$$\text{ind } SgS = \#(S\backslash SgS)\, 1_R \text{ .}$$

In spite of the fact that ind SgS is an element of the ground ring R , we sometimes assume ind SgS to be an integer. With this convention, which is always clear from the context, we have

$$\lambda(g) = \frac{\text{ind } SgS}{\text{ind } Sg^{-1}S} \text{ .}$$

(3.7) Examples. a) Let S be a normal subgroup of G . Then ι is the identity, since left and right und double cosets coincide.

b) Let (S,G) as in (3.5f) . Given $k \in \mathbb{N}$ one has

$$\text{ind } S\begin{pmatrix} 1 & 0 \\ 0 & k \end{pmatrix}S = k \text{ , } \quad \iota\left(S\begin{pmatrix} 1 & 0 \\ 0 & k \end{pmatrix}S\right) = \sum_{n=0}^{k-1} S\begin{pmatrix} 1 & n \\ 0 & k \end{pmatrix} \text{.}$$

We have a closer look at the decomposition of double cosets into right cosets.

(3.8) Lemma. *Let* (S,G) *be a Hecke pair,* $g \in G$; *and choose* $g_\nu \in G$ *such that*

$$\iota(SgS) = \sum_{\nu=1}^{d} Sg_\nu \text{ , } \quad d = \#(S\backslash SgS) \text{ .}$$

Given $s \in S$ *and* $s_\nu \in S$, $\nu = 1,\ldots,d$, *one has*

$$\iota(SgS) = \sum_{\nu=1}^{d} Ss_\nu g_\nu s .$$

Proof. The assertion follows from

$$SgS = (SgS)s = \Big(\bigcup_{\nu=1}^{d} Sg_\nu\Big)s = \bigcup_{\nu=1}^{d} Sg_\nu s = \bigcup_{\nu=1}^{d} Ss_\nu g_\nu s . \qquad \square$$

A connection between the R-modules $\mathcal{H}(S,G)$ and $\mathcal{R}(S,G)$ is given by

(3.9) Proposition. *Let* (S,G) *be a Hecke pair. Then the mapping*

$$\iota : \mathcal{H}(S,G) \longrightarrow \mathcal{R}(S,G)^S$$

is an isomorphism of the R*-modules.*

Proof. Clearly the mapping $\iota : \mathcal{H}(S,G) \longrightarrow \mathcal{R}(S,G)$ is R-linear and injective, since mutually different double cosets yield mutually different right cosets. The image is contained in $\mathcal{R}(S,G)^S$ by virtue of (3.8) .

Now let

$$U = \sum_{g:S\backslash G} u(Sg)\, Sg \in \mathcal{R}(S,G)^S .$$

Then one has

$$u(Sgs) = u(Sg) \quad \text{for all} \quad g \in G , \ s \in S$$

and therefore

$$U = \sum_{g:S\backslash G/S} u(Sg) \sum_{h:S\backslash SgS/S} Sh = \iota\Big(\sum_{g:S\backslash G/S} u(Sg)SgS\Big) . \qquad \square$$

(3.10) Exercises. a) Show that the homomorphism λ in (3.6) is trivial, i.e., $\lambda \equiv 1$, whenever S is finite.

b) Let S be a subgroup of a group G. Show that there exists a set of simultaneous representatives of the left and right cosets relative to S in SgS, $g \in G$, whenever

$$\#\big(S\backslash SgS\big) = \#\big(SgS/S\big) < \infty .$$

c) Let S_1, S_2 be two commensurable subgroups of a group G. Derive the analogue of (3.1) for (S_1,S_2)-double cosets.

d) Let S_1, S_2 be two commensurable subgroups of a group G and assume moreover that $(S_1 \cap S_2, G)$ is a Hecke pair. Show that the (S_1,S_2)-double coset space $\mathcal{H}(S_1,G,S_2)$ over R and $\mathcal{R}(S_1,G)^{S_2}$ are isomorphic R-modules.

e) Let S denote the group of integral upper triangular $n \times n$ matrices with 1's on the diagonal. Determine $\mathrm{Com}(S,GL(n;\mathbb{Q}))$.

f) Let G denote the group of bijective mappings $\pi: \mathbb{N} \longrightarrow \mathbb{N}$ and S the subgroup of all π with $\pi(1) = 1$. Determine $\mathrm{Com}(S,G)$.

4. The Product

At first we give some motivation for the product, which may look strange at first glance. Assume that $R = \mathbb{C}$ and that G acts on a certain vector space V of functions $\varphi : D \longrightarrow \mathbb{C}$ just as at the end of section 2. If (S,G) is a Hecke pair, then every $g \in G$ defines an endomorphism

$$\Phi(SgS) : V^S \longrightarrow V^S, \qquad \varphi \longmapsto \sum_{\bar{g}:S\backslash SgS} \varphi \mid \bar{g},$$

due to (3.9) combined with (2) and (3) in section 2. Now we introduce a product on $\mathcal{H}(S,G)$ such that $\Phi : \mathcal{H}(S,G) \longrightarrow \mathrm{End}\, V^S$ turns out to be a homomorphism of the algebras, i.e.,

$$\Phi(SgS) \circ \Phi(ShS) = \Phi\big(SgS \cdot ShS\big) \quad \text{for all } g,h \in G.$$

Let (S,G) be a Hecke pair and R be an arbitrary ring again. Then we can define an action of double cosets on the right coset space $\mathcal{R}(S,G)$. Given $g,h \in G$ define

$$(SgS) * (Sh) := \iota(SgS)h = \sum_{\bar{g}:S\backslash SgS} S\bar{g}h \in \mathcal{R}(S,G).$$

We have to show that the definition does not depend on the choice of representative of the right coset Sh. If we replace h by sh, $s \in S$, we obtain

$$\iota(SgS)(sh) = \big(\iota(SgS)s\big)h = \iota(SgS)h$$

in view of $\iota(SgS) \in \mathcal{R}(S,G)^S$ due to (3.9). Hence the action is well-defined.

We extend this action by linearity. Given

$$T = \sum_{g:S\backslash G/S} t(SgS)\, SgS \in \mathcal{H}(S,G),$$

$$U = \sum_{h:S\backslash G} u(Sh)\, Sh \in \mathcal{R}(S,G),$$

set

$$\begin{aligned} T * U &= \sum_{g:S\backslash G/S} \sum_{h:S\backslash G} t(SgS)\, u(Sh)\, \big[(SgS) * (Sh)\big] \\ &= \sum_{\bar{g}:S\backslash G} \sum_{h:S\backslash G} t(S\bar{g}S)\, u(Sh)\, S\bar{g}h \in \mathcal{R}(S,G). \end{aligned}$$

The decisive observation is contained in

(4.1) Proposition. *Let* (S,G) *be a Hecke pair. Given* $T \in \mathcal{H}(S,G)$ *and* $U \in \mathcal{R}(S,G)^S$ *one has*

$$T * U \in \mathcal{R}(S,G)^S .$$

Proof. Given $s \in S$ and T,U as above, we get

$$(T*U)s = \sum_{g:S\backslash G} \sum_{h:S\backslash G} t(SgS)u(Sh)\ Sghs = T*(Us) = T*U$$

because of $U \in \mathcal{R}(S,G)^S$, hence $T*U \in \mathcal{R}(S,G)^S$. □

Combining (4.1) with (3.9) we can define a product on $\mathcal{H}(S,G)$ via

(1) $$T_1 \cdot T_2 = \iota^{-1}(T_1 * \iota(T_2)) \in \mathcal{H}(S,G) \quad \text{for } T_1,T_2 \in \mathcal{H}(S,G) .$$

(4.2) Theorem. *Let* (S,G) *be a Hecke pair. Then* $\mathcal{H}(S,G)$ *endowed with the product* (1) *is an associative* R*-algebra with* $S = SeS$ *as its identity element.*

Proof. Clearly S becomes the identity element. The associative law follows from the definition of the product in $\mathcal{H}(S,G)$ and the associative law in G, i.e.,

$$\iota(T_1 \cdot (T_2 \cdot T_3)) = \iota((T_1 \cdot T_2) \cdot T_3)$$
$$= \sum_{g_1:S\backslash G} \sum_{g_2:S\backslash G} \sum_{g_3:S\backslash G} t_1(Sg_1S)\,t_2(Sg_2S)\,t_3(Sg_3S)\ Sg_1g_2g_3 .$$ □

Definition. $\mathcal{H}(S,G)$ is called the *Hecke algebra over* R *associated with* (S,G).

Hecke algebras are generalizations of group algebras as the following example shows.

(4.3) Example. Let S be a normal subgroup of a group G, i.e., $SgS = Sg = gS$ for all $g \in G$. Then all $g,h \in G$ satisfy

$$(SgS) \cdot (ShS) = SghS = Sgh .$$

Thus the Hecke algebra $\mathcal{H}(S,G)$ coincides with the group algebra $R[G/S]$ of the factor group G/S over R.

Let us have a closer look at the product. We will give several descriptions of the multiplicities. Each description will be applied later on. In some cases the arising sets (but not their orders) depend on the choice of the representatives.

(4.4) Lemma. *Let* (S,G) *be a Hecke pair. Given* $g,h \in G$ *let*

$$SgS = \bigcup_{\mu=1}^{d(g)} Sg_\mu ,\ d(g) = \#(S\backslash SgS) ,\quad ShS = \bigcup_{\nu=1}^{d(h)} Sh_\nu ,\ d(h) = \#(S\backslash ShS) .$$

Then one has

$$(2)\qquad (SgS)\cdot(ShS) = \sum_{f:S\backslash SgShS/S} \alpha(g,h;f)\, SfS\ ,$$

where

(i) $\alpha(g,h;f) = \#\{(\mu,\nu) \mid g_\mu h_\nu \in Sf\} = \#\{S\tilde{h} \mid S\tilde{h} \subset ShS\ ,\ f\tilde{h}^{-1} \in SgS\}\ ,$

(ii) $(\mathrm{ind}\ SfS)\,\alpha(g,h;f) = \#\{(\mu,\nu) \mid g_\mu h_\nu \in SfS\}\ ,$

(iii) $(\mathrm{ind}\ SfS)\,\alpha(g,h;f) = (\mathrm{ind}\ ShS)\ \#\{S\tilde{g} \mid S\tilde{g} \subset SgS\ ,\ \tilde{g}h \in SfS\}\ ,$

(iv) $(\mathrm{ind}\ SfS)(\deg Sh^{-1}S)\,\alpha(g,h;f)$

$\qquad = (\mathrm{ind}\ ShS)\ \#\{(S\tilde{g},\tilde{h}S) \mid S\tilde{g} \subset SgS\ ,\ \tilde{h}S \subset ShS\ ,\ \tilde{g}\tilde{h} \in SfS\}\ ,$

(v) $(\mathrm{ind}\ Sf^{-1}S)(\mathrm{ind}\ SgS)\,\alpha(g,h;f)$

$\qquad = (\mathrm{ind}\ Sg^{-1}S)\ \#\{(S\tilde{g},\tilde{h}S) \mid S\tilde{g} \subset SgS\ ,\ \tilde{h}S \subset ShS\ ,\ \tilde{g}\tilde{h} \in SfS\}\ .$

Proof. (i) We observe that

$$(3)\qquad \iota\big((SgS)\cdot(ShS)\big) = \sum_{\mu=1}^{d(g)} \sum_{\nu=1}^{d(h)} Sg_\mu h_\nu\ .$$

Then compute the coefficient of Sf on the right hand side.
(ii) Each right coset in SfS has the same coefficient in (3) .
(iii) Choose a set of representatives $h_\nu = hs_\nu$ with $s_\nu \in S$, $\nu = 1,\dots,d(h)$, and apply (ii) .
(iv) Use $\alpha(g,h;f) = \alpha(g,\tilde{h};f)$ for all $\tilde{h} \in ShS$ in (iii) .
(v) Assume $R = \mathbb{Z}$ without restriction. Recall from (3.6) that $\lambda : G \to \mathbb{Q}^+$ is a homomorphism, which is trivial on S . Thus we have

$$\lambda(f) = \lambda(g)\lambda(h)$$

for all $f \in SgShS$. Insert this into (iv) . □

The description (ii) especially yields

$$\mathrm{ind}\big((SgS)\cdot(ShS)\big) = \sum_{f:S\backslash SgShS/S} \alpha(g,h;f)\,\mathrm{ind}\ SfS \ = \#\{(\mu,\nu)\}$$

$$= (\mathrm{ind}\ SgS)\ (\mathrm{ind}\ ShS)\ .$$

Linear extension leads to

(4.5) Corollary. *Let* (S,G) *be a Hecke pair. Then the map*

$$\mathrm{ind}\ :\ \mathcal{H}(S,G) \to R$$

is a homomorphism of the R*-algebras.*

Next we describe the product of particular elements.

(4.6) Lemma. *Let* (S,G) *be a Hecke pair and* $g \in G$ *satisfying* $Sg = gS$. *Then one has for all* $h \in G$

(4) $$(SgS)\cdot(ShS) = SghS\ ,\ (ShS)\cdot(SgS) = ShgS\ .$$

Proof. Let $ShS = \bigcup_{\nu=1}^{d} Shs_\nu\ ,\ s_\nu \in S\ ,\ d = \text{ind } ShS$. Assume $sghs_\nu = ghs_\mu$ for some $s \in S$. We obtain $sg = g\bar{s}$ for an appropriate $\bar{s} \in S$ from $Sg = gS$. This leads to $\bar{s}hs_\nu = hs_\mu$, hence to $\nu = \mu$. Thus the $Sghs_\nu$ are d mutually disjoint right cosets in $SghS$, hence

$$(SgS)\cdot(ShS) = SghS\ .$$

On the other hand the right cosets $Shs_\nu g\ ,\ \nu = 1,\dots,d$, are mutually disjoint. Given s_ν there exists $\bar{s}_\nu \in S$ with $s_\nu g = g\bar{s}_\nu$. Thus the right cosets $Shs_\nu g = Shg\bar{s}_\nu$ are contained in the double coset $ShgS$, hence

$$(ShS)\cdot(SgS) = ShgS\ .$$ □

Note that (4) does not hold in general under the weaker assumption $\#(S\backslash SgS) = 1$.

One can strengthen (4.6) for special rings R . As usual we say that $\text{char } R = 0$, whenever $nr = 0$ for some $n \in \mathbb{Z}$ and $0 \neq r \in R$ implies $n = 0$.

(4.7) Corollary. *Let* (S,G) *be a Hecke pair und assume that* $\text{char } R = 0$. *Given* $g,h \in G$ *the following assertions are equivalent:*

(i) $$(SgS)\cdot(ShS) = SghS\ .$$

(ii) $$(\text{ind } SgS)(\text{ind } ShS) = \text{ind } SghS\ .$$

Proof. We may assume $R = \mathbb{Z}$ without restriction.

(i) $\Rightarrow$ (ii): Apply (4.5) .

(ii) $\Rightarrow$ (i): Since $SghS$ occurs in $SgShS$, we have

$$(SgS)\cdot(ShS) = SghS + \sum_{f:S\backslash G/S} \alpha_f\ SfS$$

with non-negative $\alpha_f \in \mathbb{Z}$. Application of (4.5) and (ii) yields

$$\sum_{f:S\backslash G/S} \alpha_f(\text{ind } SfS) = 0\ .$$

But $\text{ind } SfS = \#(S\backslash SfS) > 0$ implies $\alpha_f = 0$ for all f . □

For an arbitrary ring R the implication "(ii) $\Rightarrow$ (i)" does not come true any longer (cf. (5.3)).

Finally we give another description of Hecke algebras. If S is a subgroup of G, then G acts on the R-module $\mathcal{R}(S,G)$ via

$$r_g : \mathcal{R}(S,G) \to \mathcal{R}(S,G)\ ,$$
$$U \mapsto Ug\ ,$$

for $g \in G$. The set $\mathrm{End}\,\mathcal{R}(S,G)$ of endomorphisms of the R-module $\mathcal{R}(S,G)$ becomes an R-algebra. Clearly the set

$$\mathrm{End}_G\mathcal{R}(S,G) = \{\varphi \in \mathrm{End}\,\mathcal{R}(S,G) \mid \varphi \circ r_g = r_g \circ \varphi \text{ for all } g \in G\}$$

of G-endomorphisms turns out to be a subalgebra.

If (S,G) is a Hecke pair and $T \in \mathcal{H}(S,G)$ define the map

$$L(T) : \mathcal{R}(S,G) \to \mathcal{R}(S,G) ,$$
$$U \mapsto T * U .$$

Now we state the fact that L defines an algebra-action on $\mathcal{R}(S,G)$.

(4.8) Theorem. *Let* (S,G) *be a Hecke pair. Then the map*

$$L : \mathcal{H}(S,G) \to \mathrm{End}_G\mathcal{R}(S,G) ,$$
$$T \mapsto L(T) ,$$

is an isomorphism of the R*-algebras.*

Proof. Given $T \in \mathcal{H}(S,G)$, $U \in \mathcal{R}(S,G)$, and $g \in G$, one has

$$L(T) \circ r_g(U) = T * (Ug) = (T * U)g = r_g \circ L(T)(U) ,$$

hence $L(T) \in \mathrm{End}_G\mathcal{R}(S,G)$. The injectivity of L follows from

$$L(T)(S) = T * S = \iota(T)$$

by virtue of (3.9) . Given $\varphi \in \mathrm{End}_G\mathcal{R}(S,G)$, consider $U = \varphi(S) \in \mathcal{R}(S,G)$. Then

$$Us = r_s \circ \varphi(S) = \varphi \circ r_s(S) = \varphi(S) = U$$

holds for all $s \in S$, hence $U \in \mathcal{R}(S,G)^S$ and $U = \iota(T)$ for some $T \in \mathcal{H}(S,G)$ according to (3.9) . Given $g \in G$ we obtain

$$f(Sg) = \varphi \circ r_g(S) = r_g \circ \varphi(S) = r_g(U) = T * Sg = L(T)(Sg) ,$$

thus $\varphi = L(T)$. □

It is also possible to start with an arbitrary subgroup S of a group G. In this case L yields an isomorphism between $\mathcal{H}(S,\mathrm{Com}(S,G))$ and $\mathrm{End}_G\mathcal{R}(S,G)$ by slight extensions of the definitions given above.

If we restrict the summation over g to a multiplicatively closed subset M of G , which contains S , we obtain a subalgebra of the Hecke algebra denoted by $\mathcal{H}(S,M)$.

(4.9) Lemma. *Let* (S,G) *be a Hecke pair and* M *a multiplicatively closed subset of* G *containing* S *. Then* $\mathcal{H}(S,M)$ *is a subalgebra of* $\mathcal{H}(S,G)$.

We will refer to $\mathcal{H}(S,M)$ as a Hecke algebra, too.

(4.10) Exercises. a) Let S_1, S_2, S_3 be three pairwise commensurable subgroups of a group G and assume that $(S_1\cap S_2\cap S_3, G)$ is a Hecke pair. Introduce the product

$$(S_1 g S_2)\cdot(S_2 h S_3)$$

as a linear combination of (S_1,S_3)-double cosets in analogy with (2) . Derive the analogue of (4.4) . Describe the product explicitly, whenever $S_1 = \{e\}$ resp. $S_2 = \{e\}$ resp. $S_3 = \{e\}$ and all the other S_i are finite.

b) Let S_1, S_2 be commensurable subgroups of a group G and assume that $(S_1\cap S_2, G)$ is a Hecke pair. Show that $\mathcal{H}(S_1, G, S_2)$ and $\mathrm{Hom}_G(\mathcal{R}(S_2,G), \mathcal{R}(S_1,G))$ (cf. (4.8)) are isomorphic R-modules.

c) Let (S,G) be a Hecke pair and $g \in G$ satisfying $\#(S\backslash SgS) = 1$. Show that

$$(\mathrm{ind}\ SghS)\ SgS\cdot ShS = (\mathrm{ind}\ ShS)\ SghS$$

holds for all $h \in G$. Find an example with $\mathrm{ind}\ SghS \neq \mathrm{ind}\ ShS$.

d) Let (S,G) be a Hecke pair and H a subgroup of G containing S . Define

$$\pi_H : \mathcal{H}(S,G) \longrightarrow \mathcal{H}(S,H),$$
$$\sum_{g:S\backslash G/S} t(SgS)\, SgS \longmapsto \sum_{h:S\backslash H/S} t(ShS)\, ShS$$

(cf. (1.3)). Given $T_1, T_2 \in \mathcal{H}(S,G)$, $U \in \mathcal{H}(S,H)$, and $\alpha, \beta \in R$ show that

(i) $\pi_H(\alpha T_1 + \beta T_2) = \alpha\, \pi_H(T_1) + \beta\, \pi_H(T_2)$.

(ii) $\pi_H(U\cdot T_1) = U\cdot \pi_H(T_1)$, $\pi_H(T_1\cdot U) = \pi_H(T_1)\cdot U$.

(iii) U is a unit in $\mathcal{H}(S,G)$ if and only if U is a unit in $\mathcal{H}(S,H)$.

5. The Symmetric Group as an Example

In this section we discuss the simplest example of a Hecke algebra, which fails to be a group algebra. We will give an explicit description of the product.

Let S_n denote the symmetric group of n letters. In order to avoid embeddings set

$$S_n = \{\pi : \mathbb{N} \longrightarrow \mathbb{N} \mid \pi \text{ is bijective}, \ \pi(k) = k \text{ for } k > n\}.$$

Due to the definition S_{n-1} becomes a subgroup of S_n of index n . Define the sets

$$R_k = \{\pi \in S_n \mid \pi(k) = n\}, \quad L_k = \{\pi \in S_n \mid \pi(n) = k\}.$$

for $1 \le k \le n$. Hence one has $R_n = L_n = S_{n-1}$.

Claim 1. S_n *decomposes into* n *right cosets* $R_1,\ldots,R_n$ *and into* n *left cosets* $L_1,\ldots,L_n$ *relative to* S_{n-1} . *Moreover* S_{n-1} *fails to be normal in* S_n , *whenever* $n \ge 3$.

Proof. Given $\pi,\bar{\pi} \in R_k$ and $\sigma \in S_{n-1}$ one has

$$\sigma \circ \pi(k) = \sigma(n) = n, \quad \text{hence } \sigma \circ \pi \in R_k,$$
$$\bar{\pi} \circ \pi^{-1}(n) = \bar{\pi}(k) = n, \quad \text{hence } \bar{\pi} \in S_{n-1}\pi.$$

Thus $R_1,\ldots,R_n$ are mutually distinct right cosets, whose union clearly equals S_n. The left cosets are dealt with in the same way. If $n \geq 3$ the left coset L_1 fails to be a right coset, hence S_{n-1} is not normal in S_n. □

Claim 2. *There are exactly* 2 *double cosets relative to* S_{n-1} *in* S_n *for* $n \geq 2$, *namely* S_{n-1} *and*

$$\bar{S}_{n-1} = \{\pi \in S_n \mid \pi(n) \neq n\} = \{\pi \in S_n \mid \pi \notin S_{n-1}\}.$$

Proof. Given $1 \leq k,\ell \leq n$, $k \neq \ell$, let $\tau_{k,\ell} \in S_n$ denote the transposition of k and ℓ. In view of Claim 1 the set $\{\tau_{1,n}, \tau_{2,n}, \ldots, \tau_{n-1,n}, \mathrm{id}\}$ becomes a set of simultaneous representatives of the left and right cosets relative to S_{n-1} in S_n. If $n \geq 3$ and $1 < k < n$ one has $\tau_{1,k} \in S_{n-1}$ as well as

$$\tau_{1,k} \circ \tau_{k,n} \circ \tau_{1,k} = \tau_{1,n}.$$

Clearly S_{n-1} and $\bar{S}_{n-1}$ are distinct double cosets. □

We consider the Hecke algebra $\mathcal{H}(S_{n-1},S_n)$, which has S_{n-1} as its identity element.

Claim 3. $\bar{S}_{n-1} \cdot \bar{S}_{n-1} = (n-1)S_{n-1} + (n-2)\bar{S}_{n-1}$.

Proof. We have

$$\iota(\bar{S}_{n-1}) = \sum_{k=1}^{n-1} R_k = \sum_{k=1}^{n-1} S_{n-1}\tau_{k,n}.$$

Now observe that for $1 \leq k,\ell \leq n-1$

$$\tau_{k,n} \circ \tau_{\ell,n} \in \begin{cases} S_{n-1} & \text{if } k = \ell, \\ R_k & \text{if } k \neq \ell. \end{cases}$$

The right coset S_{n-1} appears $(n-1)$-times and each R_k exactly $(n-2)$-times. □

Summarizing we have the following

(5.1) Lemma. *Given* $n \geq 2$ *the Hecke algebra* $\mathcal{H}(S_{n-1},S_n)$ *is a commutative, two-dimensional* R*-algebra with* S_{n-1} *as its identity element, whose product is determined by*

$$\bar{S}_{n-1} \cdot \bar{S}_{n-1} = (n-1)S_{n-1} + (n-2)\bar{S}_{n-1}.$$

If $R = F$ is a field, we are going to describe the structure of this Hecke algebra more precisely.

(5.2) Lemma. *Let* $R = F$ *be a field and* $n \geq 2$.

a) *The units in the algebra* $\mathcal{H}(S_{n-1}, S_n)$ are given by

$$\alpha S_{n-1} + \beta \bar{S}_{n-1} ,$$

where $\alpha, \beta \in F$ *and*

$$(\alpha - \beta)(\alpha + (n-1)\beta) \neq 0 .$$

b) *If the characteristic of* F *does not divide* n , *the map*

$$\mathcal{H}(S_{n-1}, S_n) \longrightarrow F \times F ,$$
$$\alpha S_{n-1} + \beta \bar{S}_{n-1} \longmapsto (\alpha + (n-1)\beta, \alpha - \beta) ,$$

is an isomorphism of the F*-algebras.*

Proof. a) $T = \alpha S_{n-1} + \beta \bar{S}_{n-1}$ is a unit if the left multiplication by T on $\mathcal{H}(S_{n-1}, S_n)$ is bijective. The matrix of this multiplication with respect to the basis $S_{n-1}, \bar{S}_{n-1}$ is given by

$$\begin{pmatrix} \alpha & (n-1)\beta \\ \beta & \alpha + (n-2)\beta \end{pmatrix}$$

according to (5.1) and its determinant equals

$$(\alpha - \beta)(\alpha + (n-1)\beta) .$$

b) Define $T_1 := \frac{1}{n}(S_{n-1} + \bar{S}_{n-1})$, $T_2 := \frac{1}{n}((n-1)S_{n-1} - \bar{S}_{n-1})$. Now (5.1) leads to

$$T_1^2 = T_1 , T_2^2 = T_2 , T_1 \cdot T_2 = 0 , T_1 + T_2 = S_{n-1} .$$

The assertion follows from

$$\alpha S_{n-1} + \beta \bar{S}_{n-1} = (\alpha + (n-1)\beta) T_1 + (\alpha - \beta) T_2 . \qquad \square$$

If char $F | n$ one calculates

$$(S_{n-1} + \bar{S}_{n-1})^2 = n (S_{n-1} + \bar{S}_{n-1}) = 0 .$$

In this case the Hecke algebra $\mathcal{H}(S_{n-1}, S_n)$ it not isomorphic to $F \times F$, since the latter algebra does not contain any nilpotent elements.

(5.3) Exercises. a) Find an example of a Hecke algebra over a field F such that the implication "(ii) $\Rightarrow$ (i)" in (4.7) does not come true.

b) Determine the units of $\mathcal{H}(S_{n-1}, S_n)$ over $R = \mathbb{Z}$.

c) Determine all the idempotents in $\mathcal{H}(S_{n-1}, S_n)$ over a field F .

d) Determine the characteristic polynomial of an arbitrary element of the Hecke algebra $\mathcal{H}(S_{n-1}, S_n)$ over a field F .

e) Assume that F is a field with $\operatorname{char} F \mid n$. Determine the subalgebra of all the nilpotent elements $T \in \mathcal{H}(S_{n-1},S_n)$ (i.e., $T^m = 0$ for some $m \in \mathbb{N}$) .

f) Let $1, i, j, k$ be the canonical basis of the Hamiltonian quaternions over $\mathbb{R}$, i.e., $k = ij = -ji$, $i^2 = j^2 = -1$ (cf. [26] , p. 97). Set $\rho = \frac{1}{2}(1+i+j+k)$ and let G denote the multiplicative group generated by i , ρ . Show that G has the order 24 . (G is the group of units in the Hurwitz order, cf. [34] , I, §1 .) Let S denote the subgroup of G generated by ρ . Show that $\dim \mathcal{H}(S,G) = 2$ and determine the product.

6. Isomorphisms of Hecke Algebras

The basic isomorphism theorems for groups lead to isomorphisms of the associated Hecke algebras. In particular we can formulate the first and second isomorphism theorem for groups in the context of Hecke algebras.

(6.1) Theorem. *Let* (S,G) *be a Hecke pair and suppose that* $\varphi : G \longrightarrow H$ *is a homomorphism of the groups such that* $\operatorname{kernel} \varphi \subset S$. *Then* $(\varphi(S),\varphi(G))$ *is a Hecke pair, too. The induced mapping*

$$\Phi : \mathcal{H}(S,G) \to \mathcal{H}(\varphi(S),\varphi(G)) ,$$

$$T = \sum_{g:S\backslash G/S} t(SgS)\, SgS \;\mapsto\; \sum_{g:S\backslash G/S} t(SgS)\, \varphi(S)\varphi(g)\varphi(S) ,$$

is an isomorphism of the algebras and satisfies

$$\operatorname{ind} \Phi(T) = \operatorname{ind} T \quad \textit{for all} \quad T \in \mathcal{H}(S,G) .$$

Proof. Let $g \in G$ and $SgS = \bigcup_{\nu=1}^{d} Sg_\nu$, $d = \#(S\backslash SgS)$. Then

(1) $$\varphi(SgS) = \varphi(S)\varphi(g)\varphi(S) = \bigcup_{\nu=1}^{d} \varphi(S)\varphi(g_\nu)$$

holds as a disjoint union in view of $\operatorname{kernel} \varphi \subset S$. Thus $(\varphi(S),\varphi(G))$ is a Hecke pair and satisfies

$$\operatorname{ind} \Phi(SgS) = d = \operatorname{ind} SgS .$$

In view of (1) the definition of the product immediately yields that Φ is an isomorphism of the algebras. □

We apply (6.1) to the projection onto a factor group of G over a normal subgroup N . This particular case is separately formulated as

(6.2) Corollary. *Let* (S,G) *be a Hecke pair and suppose that* N *is a normal subgroup of* G *and that* N *is contained in* S . *Then* $(S/N,G/N)$ *is a Hecke pair , too. The canonical projection* $\pi : G \longrightarrow G/N$ *induces an isomorphism*

$$\Pi : \mathcal{H}(S,G) \longrightarrow \mathcal{H}(S/N,G/N)$$

of the algebras satisfying

$$\text{ind } \Pi(T) = \text{ind } T \quad \textit{for all } T \in \mathcal{H}(S,G) .$$

The corollary corresponds to the second isomorphism theorem for groups (cf. e.g. N. Jacobson [26] , p.63).

Let (S,G) be a Hecke pair. If $\varphi : G \longrightarrow H$ is an arbitrary homomorphism of the groups, then $(\varphi(S),\varphi(G))$ is always a Hecke pair. The induced map Φ is a surjective homomorphism of the R-modules, but in general fails to be a homomorphism of the algebras. As an example choose $H = \{e\}$ and let φ be the trivial homomorphism. If $S = S_{n-1}$ and $G = S_n$ as in (5.1) , we obtain

$$\Phi\left(\bar{S}_{n-1} \cdot \bar{S}_{n-1}\right) = \Phi\left((n-1)\, S_{n-1} + (n-2)\,\bar{S}_{n-1}\right) = (2n-3)\{e\} ,$$

$$\Phi\left(\bar{S}_{n-1}\right) \cdot \Phi\left(\bar{S}_{n-1}\right) = \{e\} \cdot \{e\} = \{e\} .$$

Next we give another application of (6.1) . We consider the case of direct products. In analogy with the corresponding result for group algebras (cf. [51] , p. 25) we obtain the tensor product of the Hecke algebras.

(6.3) Theorem. *Let* (S_1,G_1), (S_2,G_2) *be two Hecke pairs. Then* $(S_1 \times S_2, G_1 \times G_2)$ *is a Hecke pair, too. The Hecke algebra* $\mathcal{H}(S_1 \times S_2, G_1 \times G_2)$ *is isomorphic to the tensor product*

$$\mathcal{H}(S_1,G_1) \otimes \mathcal{H}(S_2,G_2) .$$

Proof. Given $g_1 \in G_1$ and $g_2 \in G_2$ the definition of the product yields

$$\text{ind}\left((S_1 \times S_2)(g_1,g_2)(S_1 \times S_2)\right) = \left(\text{ind } S_1 g_1 S_1\right)\left(\text{ind } S_2 g_2 S_2\right) .$$

Thus $(S_1 \times S_2, G_1 \times G_2)$ is a Hecke pair if and only if (S_1,G_1) and (S_2,G_2) are both Hecke pairs. If e_1 resp. e_2 denotes the identity element of G_1 resp. G_2 , the definition of the product yields

$$\begin{aligned}
&(S_1 \times S_2)(g_1,e_2)(S_1 \times S_2) \cdot (S_1 \times S_2)(e_1,g_2)(S_1 \times S_2) \\
= \;&(S_1 \times S_2)(e_1,g_2)(S_1 \times S_2) \cdot (S_1 \times S_2)(g_1,e_2)(S_1 \times S_2) \\
= \;&(S_1 \times S_2)(g_1,g_2)(S_1 \times S_2) .
\end{aligned}$$

Since the double cosets form a basis of the Hecke algebra, we get

$$\mathcal{H}(S_1 \times S_2, G_1 \times G_2) = \mathcal{H}(S_1 \times S_2, G_1 \times S_2) \otimes \mathcal{H}(S_1 \times S_2, S_1 \times G_2) .$$

But the Hecke algebras $\mathcal{H}(S_1 \times S_2, G_1 \times S_2)$ and $\mathcal{H}(S_1, G_1)$ as well as $\mathcal{H}(S_1 \times S_2, S_1 \times G_2)$ and $\mathcal{H}(S_2, G_2)$ are canonically isomorphic due to (6.1) applied to the projections. □

Moreover we can also express the first isomorphism theorem for groups (cf. [26], p. 64) in terms of the associated Hecke algebras. The corresponding result is due to A.N. Andrianov [4], section 3.1.3.

Let S and H be subgroups of a group G and assume that

(2) $$G = SH = \{ sh \mid s \in S,\ h \in H \}.$$

Then we define the mapping

$$\nu : \mathcal{R}(S,G) \to \mathcal{R}(S\cap H,H),$$
$$\sum_{g:S\backslash G} u(Sg)\,Sg \mapsto \sum_{h:(S\cap H)\backslash H} u(Sh)\,(S\cap H)h.$$

Clearly ν is well-defined and R-linear. In view of (2) each right coset relative to S in G possesses a representative in H. Given $h_1, h_2 \in H$ with $Sh_1 = Sh_2$ we conclude $h_1h_2^{-1} \in S\cap H$, hence $(S\cap H)h_1 = (S\cap H)h_2$. Thus we obtain

$$\nu(Sh) = (S\cap H)h \quad \text{for all } h \in H.$$

Hence $\nu : \mathcal{R}(S,G) \to \mathcal{R}(S\cap H,H)$ is an isomorphism of the R-modules. Next $(S\cap H) \subset S$ leads to

(3) $$\nu\left(\mathcal{R}(S,G)^S\right) \subset \mathcal{R}(S\cap H,H)^{S\cap H}.$$

Now additionally assume that (S,G) is a Hecke pair. Then $(S\cap H,H)$ proves to be a Hecke pair, too. In order to see this let $h \in H$ and consider mutually different right cosets $(S\cap H)h_i$, $i \in J$, in the double coset $(S\cap H)h(S\cap H)$. Then Sh_i, $i \in J$, are mutually distinct right cosets in ShS due to our considerations above. According to the Hecke condition for (S,G) now J turns out to be finite and we have

$$\#\big((S\cap H)\backslash(S\cap H)h(S\cap H)\big) \le \#\big(S\backslash ShS\big) \quad \text{for all } h \in H.$$

According to (3) and (3.9) we obtain an injective homomorphism

$$\nu : \mathcal{H}(S,G) \to \mathcal{H}(S\cap H,H)$$

of the R-modules. Given $g \in G$ we get

$$\nu(SgS) = \sum_{h:(S\cap H)\backslash((SgS)\cap H)/(S\cap H)} (S\cap H)h(S\cap H).$$

If $g,h \in H$ the definition immediately leads to

$$\nu\big(SgS * Sh\big) = \nu\big(SgS\big) * Sh.$$

Thus ν is also a homomorphism of the algebras and ν defines an embedding between the two Hecke algebras.

(6.4) Proposition. *Let* (S,G) *be a Hecke pair and* H *a subgroup of* G *satisfying*

$$G = SH = \{ sh \mid s \in S,\ h \in H \}.$$

Then $(S\cap H,H)$ *is also a Hecke pair and the mapping*

$$\nu : \mathcal{H}(S,G) \longrightarrow \mathcal{H}(S\cap H,H)$$

is an injective homomorphism of the R-*algebras. If moreover*

$$\#\big((S\cap H)\backslash(S\cap H)h(S\cap H)\big) = \#\big(S\backslash ShS\big) \quad \textit{for all } h\in H , \tag{4}$$

then ν *is an isomorphism.*

Proof. Let (2) hold and let an element $h\in H$ be given. If $d = \#\big((S\cap H)\backslash(S\cap H)h(S\cap H)\big)$ and $(S\cap H)h_i$, $1\le i\le d$, are the right cosets in $(S\cap H)h(S\cap H)$, we conclude that Sh_i , $1\le i\le d$, are d mutually distinct right cosets in ShS . Then (4) yields

$$ShS = \bigcup_{i=1}^{d} Sh_i .$$

Hence we get the surjectivity of ν by virtue of

$$\nu(ShS) = (S\cap H)h(S\cap H) . \qquad \square$$

If char R = 0 , one can easily demonstrate that the condition (4) is also necessary for ν being onto.

There is another canonical way of constructing automorphisms of a Hecke algebra.

(6.5) Lemma. *Let* (S,G) *be a Hecke pair. Assume that* $R = F$ *is a field as well as that* $\lambda : G \longrightarrow F\backslash\{0\}$ *is a homomorphism of the groups satisfying* $S \subset \text{kernel}\,\lambda$. *Then the map*

$$\Lambda : \mathcal{H}(S,G) \longrightarrow \mathcal{H}(S,G) ,$$
$$\sum_{g:S\backslash G/S} t(SgS)\, SgS \longmapsto \sum_{g:S\backslash G/S} \lambda(g)\, t(SgS)\, SgS ,$$

is an automorphism of the Hecke algebra.

Proof. Clearly Λ is well-defined and F-linear. Given $g,h\in G$ and $f\in SgShS$ we get $\lambda(f) = \lambda(g)\lambda(h)$ from $S\subset \text{kernel}\,\lambda$. Then the multiplicativity of Λ follows from the definition of the product. Λ is bijective, because the inverse map is induced by λ^{-1} .
$\square$

Finally we can derive an embedding of the Hecke algebra $\mathcal{H}(S,G)$ into the group algebra of G , whenever S is a finite group.

(6.6) Theorem. *Let* S *be a finite subgroup of a group* G . *Suppose that* $R = F$ *is a field and that* char $F \nmid \#(S)$.

a)
$$e_S := \frac{1}{\#(S)} \sum_{s\in S} s$$

is an idempotent in the group algebra $F[G]$. *One has*

$$e_S\, g\, e_S = \frac{1}{\#(S)\cdot \text{ind}\, SgS} \sum_{h\in SgS} h \tag{5}$$

for all $g \in G$ *and* $e_S F[G] e_S$ *becomes a subalgebra of* $F[G]$ *with* e_S *as its identity element.*

b) *The mapping*

$$\Phi : \mathcal{H}(S,G) \to e_S F[G] e_S ,$$
$$\sum_{g:S\backslash G/S} t(SgS)\, SgS \mapsto \frac{1}{\#(S)} \sum_{g\in G} t(SgS)\, g ,$$

is an isomorphism of the algebras.

Proof. a) $e_S{}^2 = e_S$ follows from the definiton of the product. Thus e_S equals the identity element of $e_S F[G] e_S$. Moreover (5) follows from (3.1). Note that $\#(S\backslash SgS)$ is a divisor of $\#(S)$.

b) The image of Φ equals $e_S F[G] e_S$ due to (5). Clearly Φ is F-linear and injective. Hence it suffices to prove the multiplicativity of Φ for double cosets. Let $g,h \in G$ and

$$\iota(SgS) = \sum_{\mu=1}^{d(g)} Sg_\mu , \quad \iota(ShS) = \sum_{\nu=1}^{d(h)} Sh_\nu .$$

According to (4.4) we have

$$(SgS)\cdot(ShS) = \sum_{f:S\backslash SgShS/S} \alpha(g,h;f)\, SfS ,$$

where

$$\alpha(g,h;f) = \#\left\{\nu \mid 1 \le \nu \le d(h) ,\ fh_\nu^{-1} \in SgS \right\} .$$

On the other hand we calculate for $f \in SgShS$

$$\#\left\{(\tilde{g},\tilde{h}) \mid \tilde{g} \in SgS ,\ \tilde{h} \in ShS ,\ \tilde{g}\tilde{h} = f\right\}$$
$$= \#(S)\cdot \#\left\{(\tilde{g},h_\nu) \mid \tilde{g} \in SgS ,\ \nu = 1,\dots,d(h) ,\ \tilde{g} = f h_\nu^{-1}\right\}$$
$$= \#(S)\cdot \alpha(g,h;f) .$$

□

Clearly $e_S F[G] e_S$ equals the set of linear combinations, where the coefficients of those elements, which belong to the same double coset relative to S, coincide. In algebraic terminology the Hecke algebra can be identified with the Peirce-1-space of the group algebra $F[G]$ in the Peirce decomposition with respect to the idempotent e_S (cf. [11], I,§ 12).

We rewrite (6.6) in terms of mappings and abtain the F-vector space

$$\mathcal{A}(S,G) := \left\{ \varphi : G \to F \mid \varphi(s_1 g s_2) = \varphi(g) \text{ for all } s_1, s_2 \in S \text{ and } g \in G \right\}$$

of all S bi-invariant functions, which endowed with the convolution is an associative subalgebra of the group algebra $F[G]$.

(6.7) Corollary. *Let* S *be a finite subgroup of a group* G *and* F *a field satisfying* char $F \nmid \#(S)$. *Then the mapping*

$$\Psi : \mathcal{H}(S,G) \to \mathcal{A}(S,G) ,$$
$$\sum_{g:S\backslash G/S} t(SgS)\, SgS \mapsto \Psi_T ,$$

is an isomorphism of the algebras, where

$$\Psi_T(g) := \frac{t(SgS)}{\#(S)} \quad \textit{for all} \quad g \in G.$$

(6.8) Exercises. a) Let S_1, S_2 be finite subgroups of a group G and F a field satisfying char $F \nmid \#(S_1)\cdot\#(S_2)$. Show that the double coset space $\mathcal{H}(S_1,G,S_2)$ (cf. (4.10)) and $e_{S_1}F[G]\,e_{S_2}$ are isomorphic F-vector spaces.
b) Let p be a prime and

$$S = \left\{ \begin{pmatrix} a & 0 \\ 0 & d \end{pmatrix} \mid a,d \in \mathbb{Z}/p\mathbb{Z} \quad , \; ad \neq 0 \right\} , \quad S^* = S \cap SL(2;\mathbb{Z}/p\mathbb{Z}) .$$

Are the Hecke algebras $\mathcal{H}(S,GL(2;\mathbb{Z}/p\mathbb{Z}))$ and $\mathcal{H}(S^*,SL(2;\mathbb{Z}/p\mathbb{Z}))$ isomorphic?
c) Let (S,G) be a Hecke pair and let $h \in G$ belong to the normalizer of S in G , i.e., $hS = Sh$. Show that the conjugation with h

$$G \to G ,$$
$$g \mapsto hgh^{-1} ,$$

induces an automorphism of the Hecke algebra $\mathcal{H}(S,G)$.

7. Anti-Homomorphisms

Besides homomorphisms, anti-homomorphisms also play a decisive role. Again we can prove a lifting of these anti-homomorphisms from the groups to the attached algebras. As an application we obtain a sufficient condition for the Hecke algebra to be commutative. In the sequel this result will always be used in order to demonstrate the commutativity of a Hecke algebra.

Given groups G,H a map $\varphi : G \to H$ is called an *anti-homomorphism of the groups* if

$$\varphi(g_1 g_2) = \varphi(g_2)\varphi(g_1) \quad \text{for all} \quad g_1, g_2 \in G .$$

Let $\mathcal{A},\mathcal{B}$ be R-algebras. A map $\Phi : \mathcal{A} \to \mathcal{B}$ is called an *anti-homomorphism of the algebras,* if Φ is a homomorphism of the R-modules and satisfies

$$\Phi(a_1 a_2) = \Phi(a_2)\Phi(a_1) \quad \text{for all} \quad a_1, a_2 \in \mathcal{A} .$$

The notions of anti-isomorphism and anti-automorphism are then defined in the obvious way. An anti-automorphism Φ with $\Phi \circ \Phi = \mathrm{id}$ is called an *involution.*

(7.1) Theorem. *Let* (S,G) *be a Hecke pair and suppose that* $\varphi : G \to H$ *is an anti-homomorphism of the groups such that* kernel $\varphi \subset S$. *Then* $(\varphi(S),\varphi(G))$ *is a Hecke pair, too. The induced mapping*

$$\Phi : \mathcal{H}(S,G) \to \mathcal{H}(\varphi(S),\varphi(G)) ,$$

$$T = \sum_{g:S\backslash G/S} t(SgS)SgS \quad \mapsto \quad \sum_{g:S\backslash G/S} t(SgS)\varphi(S)\varphi(g)\varphi(S) ,$$

is an anti-isomorphism of the algebras.

Proof. Given $g \in G$ the number of right cosets relative to $\varphi(S)$ in $\varphi(S)\varphi(g)\varphi(S)$ coincides with the number of left cosets relative to S in SgS , because φ is an anti-homomorphism and kernel $\varphi \subset S$ holds. Thus $(\varphi(S),\varphi(G))$ is a Hecke pair due to (3.4) and Φ clearly becomes an isomorphism of the R-modules. Given $g,h \in G$ we write the product

$$(SgS)\cdot(ShS) = \sum_{f:S\backslash SgShS/S} \alpha(g,h;f)\; SfS$$

just as in (4.4) . Use a bar for the corresponding multiplicities in the Hecke algebra $\mathcal{H}(\varphi(S),\varphi(G))$. Then it suffices to demonstrate

(1) $$\alpha(g,h;f) = \bar{\alpha}(\varphi(h),\varphi(g);\varphi(f))$$

for all $f,g,h \in G$. Therefore we may assume $R = \mathbb{Z}$. Now set

$$\beta(g,h;f) = \# \left\{ (S\tilde{g},\tilde{h}S) \mid S\tilde{g} \subset SgS , \tilde{h}S \subset ShS , \tilde{g}\tilde{h} \in SfS \right\} .$$

We use $\operatorname{ind} \varphi(S)\varphi(x)\varphi(S) = \operatorname{ind} Sx^{-1}S$ for all $x \in G$ and apply (4.4) . Inserting (iv) for $\alpha(g,h;f)$ and (v) for $\bar{\alpha}(\varphi(h),\varphi(g);\varphi(f))$ now (1) is equivalent to

$$\beta(g,h;f) = \bar{\beta}(\varphi(h),\varphi(g);\varphi(f)) .$$

But this equation follows directly from the definition by application of φ , since φ is an anti-homomorphism with kernel $\varphi \subset S$. □

If the numbers of left and right cosets in each double coset coincide, there exists a set of simultaneous representatives of the left and right cosets in each double coset (cf. (3.10)). In this case the proof of (7.1) can be simplified (cf. E. Freitag [18] , IV.1.9) .

The theorem has an obvious, but nevertheless very important consequence.

(7.2) Corollary. *Let* (S,G) *be a Hecke pair. Assume that there exists an anti-automorphism* φ *of* G *satisfying* $\varphi(S) = S$ *and*

$$S\varphi(g)S = SgS \quad \textit{for all} \quad g \in G .$$

Then the Hecke algebra $\mathcal{H}(S,G)$ *is commutative and one has*

(2) $\#(S\backslash SgS) = \#(SgS/S)$ *for all* $g \in G$.

Proof. According to (7.1) the induced anti-automorphism Φ of $\mathcal{H}(S,G)$ is the identity. Thus one has for $T_1, T_2 \in \mathcal{H}(S,G)$:

$$T_1 \cdot T_2 = \Phi(T_1 \cdot T_2) = \Phi(T_2) \cdot \Phi(T_1) = T_2 \cdot T_1 \quad .$$

Finally (2) follows from

$$\#\big(S\backslash S\varphi(g)S\big) = \#\big(SgS/S\big) \ . \qquad \square$$

Each Hecke algebra contains a canonical involution induced by the map $g \mapsto g^{-1}$.

(7.3) Example. Let (S,G) be a Hecke pair. Then the map

$$\mathcal{H}(S,G) \to \mathcal{H}(S,G) ,$$
$$T = \sum_{g:S\backslash G/S} t(SgS)SgS \mapsto \hat{T} = \sum_{g:S\backslash G/S} t(SgS)Sg^{-1}S ,$$

is an involution of $\mathcal{H}(S,G)$. We call ^ the *canonical involution* of $\mathcal{H}(S,G)$. If in particular

$$SgS = Sg^{-1}S$$

holds for all $g \in G$, then $\mathcal{H}(S,G)$ is commutative.

If G is a finite group or if the assumptions of (7.2) are satisfied, the homomorphism λ in (3.6) is trivial. In this case the identity (3) in section 3 yields

$$\operatorname{ind} \hat{T} = \operatorname{ind} T \qquad \text{for all } T \in \mathcal{H}(S,G) .$$

As a first application of (7.2) we consider the case, where G is a semi-direct product. Following N. Jacobson [26] , p. 77 , let two groups S and N be given as well as a homomorphism $\tau : S \to \operatorname{Aut} N$, $s \mapsto \tau_s$. We define a product on $N \times S$ via

$$(n,s)\cdot(\bar{n},\bar{s}) := \big(n\tau_s(\bar{n}), s\bar{s}\big) .$$

Thus we obtain a group structure on the set $N \times S$, the *semi-direct product of* N *and* S , which is denoted by $N \times_\tau S$. Due to the embedding

$$N \to N \times_\tau S \quad ,$$
$$n \mapsto (n,e) ,$$

we can regard N as a normal subgroup of $N \times_\tau S$. Moreover S can be embedded into $N \times_\tau S$ via

$$S \to N \times_\tau S \quad ,$$
$$s \mapsto (e,s) .$$

(7.4) Proposition. *Let* $G = N \times_\tau S$ *be the semi-direct product of* N *and* S . *Then* (S,G) *is a Hecke pair if and only if each* $\tau(S)$*-orbit*

$$\{\tau_s(n) \mid s \in S\}\ ,\quad n \in N\ ,$$

is finite. In this case $\mathcal{H}(S,G)$ *is commutative, whenever* N *is abelian.*

Proof. Due to the definition of the product each right coset relative to S contains a unique representative of the form

$$(n,e)\ ,\ n \in N\ .$$

A simple calculation therefore yields that the double coset $S(n,e)S$, $n \in N$, is the disjoint union of the right cosets

$$S(m,e)\quad ,\quad m \in \{\tau_s(n) \mid s \in S\}\ .$$

Assume that each $\tau(S)$-orbit is finite and that N is abelian. Consider the map

$$\varphi : G \longrightarrow G\ ,$$
$$(n,s) \mapsto (\tau_{s^{-1}}(n), s^{-1}) = (n^{-1},s)^{-1}\ .$$

A straightforward calculation yields that φ is an anti-automorphism of G satisfying $\varphi(S) = S$ and

$$S\varphi\big((n,e)\big)S \;=\; S(n,e)S\ .$$

Thus (7.2) completes the proof. □

Applications of the above result will be given in chapter III .

(7.5) Exercises. a) Let p be a prime and consider the subgroup

$$S = \left\{ \begin{pmatrix} a & b \\ 0 & 1 \end{pmatrix} \mid a,b \in \mathbb{Z}/p\mathbb{Z}\ ,\ a \neq 0 \right\}$$

of $G = GL(2;\mathbb{Z}/p\mathbb{Z})$. Show that there are exactly p double cosets relative to S in G . Is the Hecke algebra $\mathcal{H}\big(S, GL(2;\mathbb{Z}/p\mathbb{Z})\big)$ commutative?

b) Let p be a prime. Let S (resp. S') denote the subgroup of all upper (resp. lower) triangular matrices in $GL(2;\mathbb{Z}/p\mathbb{Z})$. Show that the Hecke algebras $\mathcal{H}\big(S,GL(2;\mathbb{Z}/p\mathbb{Z})\big)$ and $\mathcal{H}\big(S',GL(2;\mathbb{Z}/p\mathbb{Z})\big)$ are commutative and isomorphic.

c) Let p be a prime and $m,n \in \mathbb{N}$. On

$$G = Mat(m,n;\mathbb{Z}/p\mathbb{Z}) \times GL(m;\mathbb{Z}/p\mathbb{Z}) \times GL(n;\mathbb{Z}/p\mathbb{Z})$$

a product is introduced via

$$(A,U,V)\cdot(A^*,U^*,V^*) = (A+UA^*V^{-1},\ UU^*,\ VV^*)\ .$$

Show that G is a group and that

$$S = \{(0,U,V) \mid U \in GL(m;\mathbb{Z}/p\mathbb{Z})\ ,\ V \in GL(n;\mathbb{Z}/p\mathbb{Z})\}$$

is a subgroup. Verify that the Hecke algebra $\mathcal{H}(S,G)$ is commutative.

d) Determine all the involutions of the Hecke algebra $\mathcal{H}(S_{n-1},S_n)$ over a field F .

8. The Canonical Bilinear Form

The Hecke algebra contains a canonical linear form, which will be applied in the next chapter. If (S,G) is a Hecke pair and $T \in \mathcal{H}(S,G)$ define the *trace* of T by

$$\operatorname{tr}(T) := t(S) \in R ,$$

where $t(S)$ denotes the coefficient of S. Using the canonical involution $\hat{}$ in (7.3), we define a bilinear form by

$$\tau(T_1,T_2) = \operatorname{tr}\left(\hat{T}_1 \cdot T_2\right) \quad \text{for} \quad T_1,T_2 \in \mathcal{H}(S,G) .$$

We compute τ explicitly.

(8.1) Proposition. *Let* (S,G) *be a Hecke pair. Then* τ *is a symmetric bilinear form on* $\mathcal{H}(S,G)$. *Given*

$$T_i = \sum_{g:S\backslash G/S} t_i(SgS)\, SgS \in \mathcal{H}(S,G) , \; i = 1,2 ,$$

one has

$$\tau(T_1,T_2) = \operatorname{tr}\left(\hat{T}_1 \cdot T_2\right) = \sum_{g:S\backslash G/S} t_1(SgS)\, t_2(SgS)\, (\operatorname{ind} SgS) . \tag{1}$$

Proof. τ is symmetric in view of $\operatorname{tr}(T) = \operatorname{tr}(\hat{T})$ for all $T \in \mathcal{H}(S,G)$. In order to prove (1) the linearity of τ allows us to confine ourselves to the case $T_2 = ShS$ for some $h \in G$. Let $S\tilde{h} \subset ShS$ and consider the action of $\hat{T}_1$ on $S\tilde{h}$

$$\hat{T}_1 * S\tilde{h} = \sum_{g:G/S} t_1(SgS)\, Sg^{-1}\tilde{h} \tag{2}$$

Then $Sg^{-1}\tilde{h} = S$ holds if and only if $gS = \tilde{h}S$. Hence the coefficient of S in (2) is equal to

$$t_1(S\tilde{h}S) = t_1(ShS) .$$

This coefficient appears $\#(S\backslash ShS)$ times in the product $\hat{T}_1 \cdot (ShS)$. □

We call τ the *canonical bilinear form* on $\mathcal{H}(S,G)$. In particular the double cosets relative to S form an orthogonal basis with respect to τ. Obviously we get

(8.2) Corollary. *Let* (S,G) *be a Hecke pair and assume that* $R = F$ *is a field. Then the canonical bilinear form* τ *is non-degenerate if and only if*

$$\operatorname{char} F \nmid \#(S\backslash SgS) \quad \textit{for any} \quad g \in G .$$

(8.3) Remark. There is another canonical bilinear form, namely

$$\rho(T_1,T_2) := \operatorname{tr}\left(T_1 \cdot T_2\right) .$$

ρ is associative, i.e.,

$$\rho\left(T_1, T_2 \cdot T_3\right) = \rho\left(T_1 \cdot T_2, T_3\right) .$$

But ρ fails to be symmetric in general. Let $R = F$ be a field. Then ρ is symmetric if and only if

$$\#(S \backslash SgS) \equiv \#(SgS/S) \mod \operatorname{char} F \quad \text{for all} \quad g \in G .$$

Clearly ρ and τ coincide if and only if the canonical involution on $\mathcal{H}(S,G)$ is the identity.

(8.4) Exercises. a) Let S be a finite subgroup of a group G and F a field satisfying $\operatorname{char} F \nmid \#(S)$. Compare the trace on $\mathcal{H}(S,G)$ with the trace on $F[G] = \mathcal{H}(\{e\},G)$ given by the embedding Φ in (6.6) .
b) Find an example of a Hecke pair (S,G) and a field F, where ρ and τ are different.
c) Let (S,G) be a Hecke pair and H a subgoup of G containing S. Assume that $R = F$ is a field with $\operatorname{char} F = 0$. Given $T \in \mathcal{H}(S,H)$ show that T is a left resp. right zero-divisor in $\mathcal{H}(S,H)$ if and only if T is a left resp. right zero-divisor in $\mathcal{H}(S,G)$ (Hint: (4.10 d)).

Remarks

The results on group algebras in section 1 are well-known and can be found in most books on Algebra. The abstract Hecke algebra was introduced by G. Shimura [56] . The results of the sections 2,3,4,7 are basically due to G. Shimura and can be found in his book [58] . Some of the results, especially the identification of $\mathcal{H}(S,G)$ with $\mathcal{R}(S,G)^S$, are due to A.N. Andrianov, whose recent book [4] is also a good reference for the whole chapter. The abstract Hecke algebra is also dealt with by E. Freitag [18] .

The connection between Hecke algebras and representation theory of finite groups is pointed out by C.W. Curtis and I. Reiner [16] , [17] . Their point of view is the embedding of the Hecke algebra into the group algebra (cf. (6.6)). Then the results on the Hecke algebra follow from properties of the group algebra.

Chapter II

Structure Theory of Hecke Algebras for Finite Groups

The first obvious class of Hecke pairs consists of those pairs (S,G) , where G is a finite group. A generalization of Maschke's classical theorem for group algebras elucidates the structure of the attached Hecke algebras, whenever the ground ring is a field. The Hecke algebra associated with (S,G) proves to be semi-simple, if the characteristic of the ground field does not divide the order of G .

Then we will use the embedding of the Hecke algebra into the group algebra, which was described in I(6.6) . This embedding yields information about the Hecke algebra.

Next we have a closer look at Hecke algebras of small dimension. In particular every Hecke algebra of dimension ≤ 5 turns out to be commutative.

Finally we deal with the canonical involution again. Just as for group algebras the Hecke algebra over $\mathbb{R}$ contains as many information as the Hecke algebra with involution over $\mathbb{R}$.

Although some results are also valid for arbitrary rings, we will assume throughout the whole chapter that

$$R = F \text{ is a field,}$$

if it is not stated otherwise.

1. Maschke's Theorem for Hecke Algebras

We start with a slightly more general situation, namely with a finite dimensional subalgebra of a Hecke algebra, which was defined in I(4.9) . We can reduce this case to the case of a finite group.

(1.1) Proposition. *Let* (S,G) *be a Hecke pair and* M *a multiplicatively closed subset of* G *containing* S . *Suppose that* $\mathcal{H}$(S,M) *has finite dimension. Then*

a) M *is a group.*

b) *There exists a Hecke pair* $(\bar{S},\bar{G})$, *where* $\bar{G}$ *is a finite group, such that* $\mathcal{H}$(S,M) *and* $\mathcal{H}(\bar{S},\bar{G})$ *are canonically isomorphic.*

Proof. a) One has $r := \#(S\backslash M) = \dim \mathcal{R}(S,M) < \infty$. Let $M = \bigcup_{\nu=1}^{r} Sm_\nu$ and let $m \in M$. Then one has

$$\bigcup_{\nu=1}^{r} Sm_\nu m = M \quad ,$$

because the $Sm_\nu m$, $\nu = 1,\dots,r$, are r mutually distinct right cosets relative to S in M . Now $e \in M$ yields the existence of m^{-1} in M .

b) We define

$$\text{(1)} \qquad N := \bigcap_{m:S\backslash M} m^{-1}Sm \quad .$$

Then N is a subgroup of M of finite index. Given $n \in M$ then mn runs through a set of representatives of $S\backslash M$, whenever m does. Since the defintion (1) does not depend on the choice of the representatives, we get $n^{-1}Nn = N$. Thus N is normal in M and contained in S . Setting $\bar{S} = S/N$ and $\bar{G} = M/N$ we obtain the assertion from I(6.2) .

□

When dealing with finite dimensional Hecke algebras, we may therefore always assume that G is a finite group.

(1.2) Proposition. *Let S be a subgroup of a finite group G of index r . Setting*

$$T_0 := \sum_{g:\, S\backslash G/S} SgS \in \mathcal{H}(S,G)$$

all $T \in \mathcal{H}(S,G)$ satisfy

$$T_0 \cdot T = T \cdot T_0 = (\operatorname{ind} T)\, T_0 \, ,$$

in particular

$$T_0^{\,2} = r\, T_0 \, .$$

Proof. Given $h \in G$ the same arguments, which were used for (1.1) , yield

$$T_0 * Sh = \sum_{g:S\backslash G} Sgh = \sum_{f:S\backslash G} Sf = \iota(T_0) \, .$$

This leads to

$$T_0 \cdot T = (\operatorname{ind} T)\, T_0 \, .$$

Next use the canonical involution $T \mapsto \hat{T}$ in I(7.3) in order to obtain

$$\widehat{(T\cdot T_0)} = \hat{T}_0 \cdot \hat{T} = T_0 \cdot \hat{T} = (\operatorname{ind} \hat{T})\, T_0 = (\operatorname{ind} T)\, T_0$$

in view of $\hat{T}_0 = T_0$ and $\operatorname{ind} T = \operatorname{ind} \hat{T}$. Thus

$$T \cdot T_0 = (\text{ind } T)\, T_0$$

follows. Finally note that ind $T_0 = r$. □

If char $F \nmid r$, then (1.2) says that $\frac{1}{r}T_0$ is a primitive idempotent in the center of the Hecke algebra $\mathcal{H}(S,G)$.

We are able to derive a negative result on semi-simplicity. Therefore we need the

Definition. Let $\mathcal{A}$ be a finite dimensional F-algebra. $\mathcal{A}$ is called *simple*, if $\{0\}$ and $\mathcal{A}$ are the only ideals in $\mathcal{A}$. An ideal $\mathcal{D}$ of $\mathcal{A}$ is called *nilpotent* if there exists $k \in \mathbb{N}$ such that

$$a_1 \cdot \ldots \cdot a_k = 0 \quad \text{for all } a_1, \ldots, a_k \in \mathcal{D} .$$

The *radical* of $\mathcal{A}$ is the maximal nilpotent ideal in $\mathcal{A}$ and is denoted by rad $\mathcal{A}$. The algebra $\mathcal{A}$ is called *semi-simple*, if rad $\mathcal{A} = \{0\}$.

The famous Wedderburn Theorem (cf. [23]) gives a characterization of semi-simple algebras as direct sums of matrix algebras. Moreover $\mathcal{A}/\text{rad } \mathcal{A}$ is always semi-simple. If $\mathcal{A}$ is commutative, one has

$$\text{rad } \mathcal{A} = \{a \in \mathcal{A} \mid a \text{ nilpotent}\} .$$

(1.3) Lemma. *Let* S *be a subgroup of a finite group* G *of index* r . *If* F *is a field such that* char $F \mid r$, *the Hecke algebra* $\mathcal{H}(S,G)$ *over* F *fails to be semi-simple.*

Proof. In view of (1.2) the set FT_0 proves to be a non-zero nilpotent ideal in $\mathcal{H}(S,G)$. □

In order to derive a converse result, we have to compute the trace of $L(T) \in \text{End } \mathcal{R}(S,G)$ (cf. I(4.8)) and compare this with the trace of T (cf. I(8.1)).

(1.4) Lemma. *Let* S *be a subgroup of a finite group* G *of index* r . *Then all* $T \in \mathcal{H}(S,G)$ *satisfy*

$$\text{trace } L(T) = r \text{ tr}(T) .$$

Proof. Clearly trace $L(T)$ equals the sum of all the coefficients of Sh in $T * Sh$, were h runs trough a set of representatives of the right cosets relative to S in G . Given $g \in G$, $g \notin S$, we observe that for $h \in G$ the right coset Sh does not occur in the sum

$$\sum_{\bar{g}:S\backslash SgS} S\bar{g}h = SgS * Sh = L(SgS)(Sh) ,$$

because $S\bar{g}h = Sh$ implies $\bar{g} \in S$ and therefore $g \in S$. This leads to trace $L(SgS) = 0$ and to

$$\operatorname{trace} L(T) = \operatorname{trace} L\big(\operatorname{tr}(T)S\big) = r \operatorname{tr}(T) ,$$

because $L(S)$ is the identity and $\mathcal{R}(S,G)$ has the dimension r. □

We obtain the following application.

(1.5) Corollary. *Let* S *be a subgroup of a finite group* G *of index* r . *Suppose that* F *is a field with* $\operatorname{char} F \nmid r$.

a) *If* $T \in \mathcal{H}(S,G)$ *is nilpotent, then* $\operatorname{tr}(T) = 0$.

b) *If* $T \in \mathcal{H}(S,G)$ *is an idempotent, then*

$$\operatorname{tr}(T) = \big(\operatorname{rank} L(T)\big)/r .$$

Proof. a) $L(T)$ is a nilpotent endomorphism, hence $\operatorname{trace} L(T) = 0$. Apply (1.4) .
b) $L(T)$ is an idempotent in $\operatorname{End} \mathcal{R}(S,G)$. The assertion follows from (1.4) in view of

$$\operatorname{trace} L(T) = \operatorname{rank} L(T) .$$ □

Now we can prove Maschke's Theorem for Hecke algebras.

(1.6) Theorem. *Let* S *be a subgroup of a finite group* G *of index* r . *If* F *is a field satisfying*

$$\operatorname{char} F \nmid r \cdot \text{l.c.m.}\big\{\#(S\backslash SgS) \mid g \in G\big\} ,$$

then the Hecke algebra $\mathcal{H}(S,G)$ *over* F *is semi-simple.*

Proof. Assume that $\mathcal{H}(S,G)$ is not semi-simple. Then choose $T \neq 0$ in a nilpotent ideal of $\mathcal{H}(S,G)$. If we multiply T by a suitable double coset, we may assume $\operatorname{tr}(T) \neq 0$ due to I(8.1) and $\operatorname{ind} SgS \neq 0$ for all $g \in G$. Then (1.5) yields a contradiction. □

Recall from I(8.2) that $\operatorname{char} F \nmid \text{l.c.m.}\{\#(S\backslash SgS) \mid g \in G\}$ just means that the canonical bilinear form τ is non-degenerate.

In the case of a group algebra, i.e., S is normal in G , we have $\#(S\backslash SgS) = 1$ for all $g \in G$. Due to (1.3) and (1.6) the group algebra $F[G/S]$ turns out to be semi-simple if and only if

$$\operatorname{char} F \nmid r = \#(G/S) .$$

This is exactly Maschke's Theorem (cf. [23] or [51]).

In general there remains a gap. The structure of $\mathcal{H}(S,G)$ fails to be clear, whenever

$$\text{(2)} \qquad \operatorname{char} F \nmid r \ \text{ but } \ \operatorname{char} F \mid \text{l.c.m.}\big\{\#(S\backslash SgS) \mid g \in G\big\} .$$

In section I.5 we already dealt with an example of a semi-simple Hecke algebra satisfying (2) . If $n > 1$ and $\operatorname{char} F \mid (n-1) = \operatorname{ind} \bar{S}_{n-1}$, then $\mathcal{H}(S_{n-1},S_n)$ over F is semi-simple in view of I(5.2) . In particular this gives us an example, where the Hecke algebra $\mathcal{H}(S_{n-1},S_n)$ is semi-simple, whereas the group algebra $F[S_n]$ fails to be semi-simple. In chapter III we will give examples of non-semi-simple Hecke algebras satisfying (2) .

If the Hecke algebra fails to be semi-simple, the radical always depends on S , G, and F . In some cases of small dimension we can describe the radical explicitly.

(1.7) Remark. Let (S,G) be a Hecke pair with $1 < \dim \mathcal{H}(S;G) < \infty$. Then (1.2) yields that FT_0 is a non-trivial ideal in $\mathcal{H}(S,G)$. Thus $\mathcal{H}(S,G)$ always fails to be a simple algebra. Moreover $\mathcal{H}(S,G)$ always contains zero-divisors, since (1.2) implies

$$T_0 \cdot (T_0 - rS) = (T_0 - rS) \cdot T_0 = T_0^2 - rT_0 = 0 .$$

In the following chapters we will find examples of infinite dimensional Hecke algebras without zero-divisors.

(1.8) Exercises. Let S be a subgroup of a finite group G .

a) Show that S is the only idempotent in the Hecke algebra $\mathcal{H}(S,G)$ over $\mathbb{Z}$.

b) Show that each element of the Hecke algebra $\mathcal{H}(S,G)$ over $\mathbb{Z}$ is integral over $\mathbb{Z}$, i.e., root of a monic polynomial in $\mathbb{Z}[X]$ (cf. [42] , chap. IX) .

c) Let N be a normal subgroup of G containing S . Assume that $\varkappa,\lambda : G/N \to F\setminus\{0\}$ are homomorphisms of the groups. Show that

$$T_\varkappa := \sum_{g:S\backslash G/S} \varkappa(Ng)\, SgS \in \mathcal{H}(S,G)$$

is well-defined. Given $T \in \mathcal{H}(S,G)$ demonstrate

(i) $$T_\varkappa \cdot T = T \cdot T_\varkappa = \operatorname{tr}(T_\varkappa \cdot T)\, T_\varkappa .$$

(ii) $$T_\varkappa \cdot T_\lambda = \begin{cases} [G:S]\, T_\varkappa & \text{if } \varkappa = \lambda , \\ 0 & \text{if } \varkappa \neq \lambda . \end{cases}$$

d) Assume that N is a normal subgroup of G containing S . Show that

$$T := \sum_{g:S\backslash N/S} SgS$$

belongs to the center of $\mathcal{H}(S,G)$.

e) Let $n > 1$. Does there exist a Hecke pair (S,G) such that the Hecke algebra $\mathcal{H}(S,G)$ over $\mathbb{C}$ is isomorphic to $\mathrm{Mat}(n;\mathbb{C})$?

2. The Use of Group Algebras

In this section we apply the embedding I(6.6) of the Hecke algebra into the group algebra in order to derive results on the attached Hecke algebras.

Recall the definition of Φ from I(6.6) . If S is a finite subgroup of a group G and F a field with $\operatorname{char} F \nmid \#(S)$ one obtains an embedding

$$\Phi : \mathcal{H}(S,G) \longrightarrow F[G] ,$$
$$\sum_{g:S\backslash G/S} t(SgS)\, SgS \mapsto \frac{1}{\#(S)} \sum_{g\in G} t(SgS)\, g .$$

First of all we derive a weaker version of Maschke's Theorem, which is for instance given by N. Iwahori [24] or N. Bourbaki [9] , §2 , Ex. 22 . A combination of I(6.6) and Maschke's Theorem for group algebras (e.g. [51] , II.4.2) immediately yields

(2.1) Corollary. *Let* S *be a subgroup of a finite group* G . *The Hecke algebra* $\mathcal{H}(S,G)$ *over* F *is semi-simple, whenever*

$$\operatorname{char} F \nmid \#(G) .$$

Due to I(3.1) the quantities $\#(S\backslash SgS)$ are always divisors of $\#(S)$. Hence (2.1) also proves to be a consequence of (1.6) .

Next we have a look at the simple factors of the Hecke algebra. If $\mathcal{A}$ is a semi-simple F-algebra, an idempotent $c \in \mathcal{A}$ is called *primitive*, if $\mathcal{A}c$ is a simple $\mathcal{A}$-left module. This is equivalent to $c\mathcal{A}c$ being an extension (skew-) field of F .

(2.2) Lemma. *Let* S *be a subgroup of a finite group* G *and* F *a field satisfying* $\operatorname{char} F \nmid \#(G)$. *An idempotent* $T \in \mathcal{H}(S,G)$ *is primitive if and only if* $\Phi(T)$ *is a primitive idempotent in* F[G] .

Proof. Since e_S is the identity element in $\Phi(\mathcal{H}(S,G))$, one has

$$\Phi(T)\, e_S = e_S\, \Phi(T) = \Phi(T) .$$

In view of I(6.6) this leads to

$$\Phi\big(T\cdot\mathcal{H}(S,G)\cdot T\big) = \Phi(T)\, e_S F[G]\, e_S\, \Phi(T) = \Phi(T)\, F[G]\, \Phi(T) . \qquad \square$$

If $\mathcal{A}$ is a simple algebra and c an idempotent in $\mathcal{A}$, then $c\mathcal{A}c$ proves to be simple, too. Thus a combination of I(6.6) and the corresponding result for group algebras (cf. [51] , II.4.4) yields

(2.3) Proposition. *Let* S *be a subgroup of a finite group* G *and* F *an algebraically closed field with* $\operatorname{char} F = 0$. *Then the number of simple factors in the Wedderburn decomposition of the Hecke algebra* $\mathcal{H}(S,G)$ *over* F *does not exceed the number of conjugacy classes in* G .

The same idea can be used in order to describe the center of the Hecke algebra.

(2.4) Theorem. *Let* S *be a subgroup of a finite group* G *and* F *a field satisfying* $\operatorname{char} F \nmid \#(G)$. *If* $\mathcal{C}(G)$ *denotes the set of all conjugacy classes in* G , *the center of the Hecke algebra* $\mathcal{H}(S,G)$ *is spanned by the elements*

$$T_C = \sum_{g\in C} \frac{1}{\operatorname{ind} SgS} SgS \quad , \quad C \in \mathcal{C}(G) .$$

Proof. If $\mathcal{A} = \mathrm{Mat}(n;K)$, where K is a (skew-) field extension of F, and c is an idempotent in $\mathcal{A}$, then $U \in GL(n;K)$ exists such that

$$c = U\begin{pmatrix} I & 0 \\ 0 & 0\end{pmatrix}U^{-1} , \; I \text{ the } r\times r \text{ identity matrix and } r = \operatorname{rank} c .$$

Hence we obtain

$$\operatorname{cent}\left(c\mathcal{A}c\right) = c\left(\operatorname{cent}\mathcal{A}\right) = c\left((\operatorname{cent} K)\, I^{(n)}\right)c .$$

Since $F[G]$ is semi-simple, the Wedderburn decomposition and I(6.6) lead to

$$\Phi\left(\operatorname{cent}\mathcal{H}(S,G)\right) = e_S\left(\operatorname{cent} F[G]\right)e_S .$$

Due to I(1.2) cent $F[G]$ is spanned by the elements

$$U_C = \sum_{g\in C} g \quad , \quad C \in \mathcal{C}(G) .$$

But I(6.6) implies

$$e_S U_C e_S = \frac{1}{\#(S)} \sum_{g\in C} \frac{1}{\operatorname{ind} SgS} SgS = \Phi\left(T_C\right) , \quad C \in \mathcal{C}(G) . \qquad \square$$

Note that the T_C, $C \in \mathcal{C}(G)$, may be linearly dependent, whereas the U_C, $C \in \mathcal{C}(G)$, are linearly independent in $F[G]$. The reason is that e_S may annihilate simple factors of $F[G]$.

We can also describe results on idempotents, whenever G is infinite, but S is finite.

(2.5) Lemma. *Let* S *be a finite subgroup of a group* G *and* F *a field. Assume that* $T \neq S$ *is an idempotent in the Hecke algebra* $\mathcal{H}(S,G)$ *over* F.

a) *If* $\operatorname{char} F \nmid \#(S)$, *then* $\operatorname{tr}(T)$ *is contained in the prime field of* F.

b) *If* F *is an extension field of* $\mathbb{Q}$, *then one has*

$$\operatorname{tr}(T) \in \mathbb{Q} \quad \textit{and} \quad 0 < \operatorname{tr}(T) < 1 .$$

Proof. According to I(6.6) $\Phi(T)$ is an idempotent in $F[G]$ satisfying

$$\operatorname{tr}\Phi(T) = \frac{1}{\#(S)} \operatorname{tr}(T) .$$

a) Apply [51], II.3.5.

b) [51], II.1.8, yields $\operatorname{tr}(T) > 0$. But $S - T$ is another idempotent, hence $\operatorname{tr}(T) < 1$ follows. $\square$

Let $\mathcal{A}$ be an F-algebra. An element $a \in \mathcal{A}$ is called *algebraic over* F, if a is the root of a polynomial $0 \neq f(X) \in F[X]$. The algebra $\mathcal{A}$ is called *algebraic over* F, if every element of $\mathcal{A}$ is algebraic over F. A group G is called *locally finite*, if every finite subset of G generates a finite subgroup.

(2.6) Proposition. *Let* S *be a finite subgroup of a group* G.
a) *If* G *is locally finite, then the Hecke algebra* $\mathcal{H}(S,G)$ *is algebraic over* F.
b) *If* $\operatorname{char} F = 0$ *and the Hecke algebra* $\mathcal{H}(S,G)$ *is algebraic over* F, *then* G *is locally finite.*

Proof. a) Let $0 \neq T \in \mathcal{H}(S,G)$ be given in the standard form. Consider the finite subset

$$\{ g \in G \mid t(SgS) \neq 0 \} \cup S$$

of G, which therefore generates a finite subgroup H. Thus T belongs to the Hecke algebra $\mathcal{H}(S,H)$, which has finite dimension. Hence T is algebraic over F.
b) Let X be a finite subset of G. We obtain a finite set

$$Y = \left\{ s_1 x s_2 \,,\, s_1 x^{-1} s_2 \mid s_1, s_2 \in S \,,\, x \in X \right\},$$

which consists of full double cosets relative to S. The element

$$T = \#(S) \sum_{g:S\backslash Y/S} SgS \in \mathcal{H}(S,G)$$

is algebraic. Hence

$$\Phi(T) = \sum_{g \in Y} g$$

is algebraic in $F[G]$. Now the proof of [51], II.3.11, implies that Y generates a finite subgroup of G. Thus G is locally finite. □

(2.7) Exercises. a) Find an example of a Hecke pair (S,G), where G is finite, such that the number of simple factors of $\mathcal{H}(S,G)$ over $\mathbb{C}$ is less than the number of conjugacy classes in G. Show that then the elements T_C, $C \in \mathcal{C}(G)$, in (2.4) are linearly dependent.
b) Show that

$$\#(S)\, T_0 = \sum_{C \in \mathcal{C}(G)} T_C$$

holds under the assumptions of (2.4) and (1.2).
c) Let S be a finite subgroup of a group G and F a field with $\operatorname{char} F = 0$. Show that the Hecke algebra $\mathcal{H}(S,G)$ over F is von Neumann finite, i.e., $T_1 \cdot T_2 = S$ implies $T_2 \cdot T_1 = S$.
d) Let S be a finite subgroup of a group G. Show that the Hecke algebra $\mathcal{H}(S,G)$ over $\mathbb{Z}$ is integral (i.e., consists of integral elements only, cf. (1.8)) if and only if G is locally finite.
e) Let (S,G) be a Hecke pair and consider the Hecke algebra $\mathcal{H}(S,G)$ over $F = \mathbb{Q}$. Given $g \in G$, $g \notin S$, show that $S \cup SgS$ is a subgroup of G if and only if the element $T := S + SgS \in \mathcal{H}(S,G)$ satisfies

$$T^2 = (\operatorname{ind} T)\, T .$$

3. Hecke Algebras of Small Dimension

In this section we derive some results for Hecke algebras of small dimension. Especially all Hecke algebras of dimension ≤ 5 turn out to be commutative.

The case of dimension 1 is obvious. So we use (1.2) in order to generalize section I.5 to arbitrary two-dimensional Hecke algebras.

(3.1) Lemma. *Let* S *be a subgroup of a finite group* G *of index* r *and suppose that* $\dim \mathcal{H}(S,G) = 2$ *, i.e.,*

$$G = S \cup SgS \ , \quad g \notin S \ .$$

Then one has

a) $$(SgS)^2 = (r-1)\,S + (r-2)\,SgS \ .$$

b) *The units in* $\mathcal{H}(S,G)$ *are given by* $\alpha S + \beta SgS$ *, where* $\alpha,\beta \in F$ *satisfy*

$$(\alpha - \beta)(\alpha + (r-1)\beta) \neq 0 \ .$$

c) *If* $\operatorname{char} F \nmid r$ *, the map*

$$\mathcal{H}(S,G) \longrightarrow F \times F \ ,$$
$$\alpha S + \beta SgS \longmapsto (\alpha + (r-1)\beta \, , \alpha - \beta)$$

is an isomorphism of the algebras.

d) *If* $\operatorname{char} F \mid r$ *, then*

$$\operatorname{rad} \mathcal{H}(S,G) = F(S + SgS)$$

holds.

Proof. a) In view of $T_0 = S + SgS$ the assertion follows from (1.2) , because S is the identity element.

b),c),d) By virtue of a) one can apply the same arguments as for the proof of I(5.2) .

□

The case of three-dimensional Hecke algebras is of course more complicated. The simplest case is the one, where the canonical involution $\hat{}$ in I(7.3) fails to be the identity. An example of this type is the group algebra of a group of order 3 .

(3.2) Lemma. *Let* S *be a subgroup of a finite group* G *of index* r *. Assume that the Hecke algebra* $\mathcal{H} = \mathcal{H}(S,G)$ *has the dimension* 3 *and consists of the double cosets* S, D_1, D_2 *. Suppose that the involution* $\hat{}$ *is not the identity, i.e.,* $\hat{D}_1 = D_2$ *. Then* $\mathcal{H}$ *is commutative,* $r \equiv 3 \bmod 4$ *, and one has*

$$D_1^2 = \tfrac{r-3}{4} D_1 + \tfrac{r+1}{4} D_2 \ , \ D_2^2 = \tfrac{r+1}{4} D_1 + \tfrac{r-3}{4} D_2 \ , \ D_1 \cdot D_2 = \tfrac{r-1}{2} S + \tfrac{r-3}{4} D_1 + \tfrac{r-3}{4} D_2 \ .$$

Let $\bar{F}$ *be a splitting field of the polynomial* $X^2 + X + \frac{r+1}{4} \in F[X]$.

(i) $\mathcal{H}$ *is isomorphic to* $F \times F \times F$, *whenever*

$$\operatorname{char} F \nmid r \quad \textit{and} \quad \bar{F} = F .$$

(ii) $\mathcal{H}$ *is isomorphic to* $F \times \bar{F}$, *whenever*

$$\operatorname{char} F \nmid r \quad \textit{and} \quad \bar{F} \neq F .$$

(iii) *If* $\operatorname{char} F \mid r$, *one has*

$$\operatorname{rad} \mathcal{H} = F\, T_0 + F\,(D_1 - D_2) .$$

Proof. At first assume $F = \mathbb{Q}$. Set $d = \#(S\backslash D_1) = \#(S\backslash D_2)$ due to $\hat{D}_1 = D_2$, therefore $r = 2d+1$. I(8.1) and $\hat{D}_1 = D_2$ yield $\operatorname{tr}(D_1{}^2) = \operatorname{tr}(D_2{}^2) = 0$. Hence we get

$$D_1{}^2 = \alpha D_1 + \beta D_2 \ , \quad D_2{}^2 = \beta D_1 + \alpha D_2$$

with some $\alpha, \beta \in \mathbb{Z}$. Application of the homomorphism ind (cf. I(4.5)) yields

$$d^2 = \alpha d + \beta d \ , \quad \beta = d - \alpha \ .$$

Then (1.2) implies

$$\begin{aligned} D_1 \cdot D_2 = D_2 \cdot D_1 &= D_1 \cdot T_0 - D_1 - D_1{}^2 = dS + (d-1-\alpha) D_1 + \alpha D_2 \\ &= D_2 \cdot T_0 - D_2 - D_2{}^2 = dS + \alpha D_1 + (d-1-\alpha) D_2 \ . \end{aligned}$$

This leads to $\alpha = \frac{d-1}{2} \in \mathbb{Z}$. Thus d is odd and the first part is proved.

Now consider the case of an arbitrary ground field F . Let $\operatorname{char} F \mid \frac{r+1}{4}$. In this case $-T_0$, $S + D_1$, $S + D_2$ is a complete system of orthogonal idempotents. Thus $\mathcal{H} \cong F \times F \times F$ follows.

Now assume that $\operatorname{char} F \nmid \frac{r+1}{4}$. If we consider the left multiplication by D_1 with respect to the basis S, D_1, D_2, we obtain

$$\mathcal{H} = F[D_1] \cong F[A] \ , \quad A = \begin{pmatrix} 0 & 0 & \frac{r-1}{2} \\ 1 & \frac{r-3}{4} & \frac{r-3}{4} \\ 0 & \frac{r+1}{4} & \frac{r-3}{4} \end{pmatrix} .$$

The characteristic polynomial of A equals

$$\left(X - \tfrac{r-1}{2}\right)\left(X^2 + X + \tfrac{r+1}{4}\right) .$$

In view of $\operatorname{char} F \nmid \frac{r+1}{4}$, the roots are mutually distinct, whenever $\operatorname{char} F \nmid r$. This leads to (i) , (ii) .

If $\operatorname{char} F \mid r$, one easily checks that besides T_0 also $D_1 - D_2$ becomes nilpotent. Hence we conclude

$$\operatorname{rad}\mathcal{H} = FT_0 + F(D_1 - D_2)\ . \qquad \square$$

The situation is not that simple if the canonical involution $\hat{\ }$ is the identity. In this case analogous arguments yield

(3.3) Proposition. *Let* (S,G) *be a Hecke pair. Assume that* $\mathcal{H}(S,G)$ *has the dimension* 3 *and consists of the double cosets* S, D_1, D_2 satisfying $\hat{D}_1 = D_1$, $\hat{D}_2 = D_2$. *Define* $d_1 = \#(S\backslash D_1)$, $d_2 = \#(S\backslash D_2)$. *Then there exists* $\lambda \in \mathbb{N}_0$ *such that the product in* $\mathcal{H}(S,G)$ *is given by*

$$D_1^{\,2} = d_1 S + (d_1 - 1 - \lambda\frac{d_2}{d_1})D_1 + \lambda D_2\ ,$$

$$D_2^{\,2} = d_2 S + (d_2 - \lambda\frac{d_2}{d_1})\,D_1 + (d_2 - d_1 - 1 + \lambda)\,D_2\ ,$$

$$D_1\cdot D_2 = D_2\cdot D_1 = \lambda\frac{d_2}{d_1}\,D_1 + (d_1 - \lambda)\,D_2\ .$$

There are also some restrictions on λ , since the coefficients in (3.3) are non-negative integers. This case cannot occur as a group algebra. An actual example of this type will be given in (4.1) .

In particular we have already proved that Hecke algebras of dimension ≤ 3 are commutative. As a more general result we obtain

(3.4) Theorem. *Let* (S,G) *be a Hecke pair. The Hecke algebra* $\mathcal{H}(S,G)$ *is commutative, whenever*

$$\dim \mathcal{H}(S,G) \leq 5\ .$$

Proof. Without restriction assume $F = \mathbb{C}$ and that G is a finite group. Then T_0 belongs to the center of $\mathcal{H}(S,G)$ according to (1.2) . If $S \neq G$ the center of $\mathcal{H}(S,G)$ has therefore at least the dimension 2 . Thus (1.6) and Wedderburn's Theorem (cf. [23]) yield the commutativity of $\mathcal{H}(S,G)$, whenever $\dim \mathcal{H}(S,G) \leq 4$.

Now assume $\dim \mathcal{H}(S,G) = 5$ and that $\mathcal{H}(S,G)$ fails to be commutative. Due to (1.6) and Wedderburn's Theorem we conclude that $\mathcal{H}(S,G)$ is isomorphic to

$$\mathbb{C} \times \operatorname{Mat}(2;\mathbb{C})\ .$$

The center of $\mathcal{H}(S,G)$ is equal to $\mathbb{C}S + \mathbb{C}T_0$. If $r = [G:S]$ we conclude from (1.2) that $\frac{1}{r}T_0$ and $S - \frac{1}{r}T_0$ are orthogonal idempotents in the center of the Hecke algebra. Now $T_0\cdot\mathcal{H}(S,G) = \mathbb{C}T_0$ implies that

$$\mathcal{A} := \left(S - \tfrac{1}{r}T_0\right)\cdot\mathcal{H}(S,G)$$

is a subalgebra of $\mathcal{H}(S,G)$, which is isomorphic to $\operatorname{Mat}(2;\mathbb{C})$. Due to I(7.3) there exists a double coset D_1 in $\mathcal{H}(S,G)$ with $D_2 := \hat{D}_1 \neq D_1$. Denote the other double cosets in $\mathcal{H}(S,G)$ by $D_0 = S$, D_3 , and D_4 . If $d = \operatorname{ind} D_1 = \operatorname{ind} D_2$ we conclude from (1.2) that

$$(S - \tfrac{1}{r}T_0)\cdot D_1 = D_1 - \tfrac{d}{r}T_0 \ , \quad (S - \tfrac{1}{r}T_0)\cdot D_2 = D_2 - \tfrac{d}{r}T_0 \ .$$

Thus

$$D_1 - D_2 \in \mathcal{A}$$

follows. Because of the isomorphy of $\mathcal{A}$ and $\mathrm{Mat}(2;\mathbb{C})$, there exist $\alpha, \beta \in \mathbb{C}$ such that

$$(D_1 - D_2)^2 = \alpha(D_1 - D_2) + \beta(S - \tfrac{1}{r}T_0) \ ,$$

due to the Cayley-Hamilton-Theorem, because $S - \frac{1}{r}T_0$ is the identity element of $\mathcal{A}$. Next

$$\mathrm{tr}\big((D_1 - D_2)^2\big) = -2d$$

follows from $\hat{D}_1 = D_2$ and I(8.1). Now $T_0 = S + D_1 + D_2 + D_3 + D_4$ yields

$$\beta = \frac{-2dr}{r-1} \ .$$

Thus we have

$$(D_1 - D_2)^2 = -2dS + (\alpha + \tfrac{2d}{r-1})D_1 + (-\alpha + \tfrac{2d}{r-1})D_2 + \tfrac{2d}{r-1}(D_3 + D_4) \ .$$

By the definition of the product, the coefficients on the right hand side are integers. Hence $r-1$ is a divisor of $2d$. But

$$r = 1 + 2d + \mathrm{ind}\, D_3 + \mathrm{ind}\, D_4$$

yields $1 < 2d < r-1$ and therefore a contradiction. □

The group algebra $F[S_3]$ is an example of a non-commutative Hecke algebra of dimension 6.

(3.5) Exercises. a) Let the assumptions of (3.2) resp. (3.3) hold. Determine the exact conditions on F, under which D_1 resp. D_2 is a unit in $\mathcal{H}(S,G)$.

b) Let (S,G) satisfy the assumptions of (3.3) and set $r := [G:S]$. Show that

$$\mathcal{A} := \{ T \in \mathcal{H}(S,G) \mid \hat{T} = T \}$$

is a subalgebra of $\mathcal{H}(S,G)$, which is isomorphic to $\mathcal{H}(S_{r-1}, S_r)$ (cf. I.5).

c) Let (S,G) be a Hecke pair such that $\dim \mathcal{H}(S,G) = 3$ and assume that there exists a proper subgroup N of G, which contains S properly. Show that $\hat{T} = T$ holds for all $T \in \mathcal{H}(S,G)$. If F is a field with $\mathrm{char}\, F \nmid [G:S]$, then prove that $\mathcal{H}(S,G)$ is isomorphic to $F \times F \times F$. Determine $\mathrm{rad}\,\mathcal{H}(S,G)$, if $\mathrm{char}\, F \mid [G:S]$. An example of this type will be provided in III(1.1) with $n = 4$.

d) Let (S,G) be a Hecke pair such that $\dim \mathcal{H}(S,G) = 3$. Assume that the normalizer N of S in G, i.e., $N = \{ h \in G \mid hS = Sh \}$, contains S properly. Determine the exact conditions on F , under which $\mathcal{H}(S,G)$ is semi-simple. Determine the radical in the non-semi-simple cases.

e) Let the assumptions of (3.2) hold and consider the Hecke algebra $\mathcal{H}$ over $\mathbb{Z}$. Show that there exists an element $T \in \mathcal{H}$ satisfying $\mathcal{H} = \mathbb{Z}[T]$ if and only if $r = 3$, where $\mathcal{H}$ is isomorphic to $\mathbb{Z}[U]$, where U is the additive group $\mathbb{Z}/3\mathbb{Z}$.

f) Let (S,G) be a Hecke pair such that $\dim \mathcal{H}(S,G) = 4$. Let S, D_1, D_2, D_3 be the double cosets relative to S and assume that $S \cup D_i$ is a subgroup of G for $i = 1,2,3$. (An example of this type is the group algebra of $\mathbb{Z}/2\mathbb{Z} \times \mathbb{Z}/2\mathbb{Z}$.) Demonstrate

(i) The canonical involution on $\mathcal{H}(S,G)$ is the identity.

(ii) $\operatorname{ind} D_1 = \operatorname{ind} D_2 = \operatorname{ind} D_3 =: d$.

(iii) $D_1 \cdot D_2 = d D_3$, $D_1 \cdot D_3 = d D_2$, $D_2 \cdot D_3 = d D_1$, $D_i^2 = d S + (d-1) D_i$, $i = 1,2,3$.

(iv) $\mathcal{H}(S,G)$ over a field F is semi-simple if and only if $\operatorname{char} F \nmid (3d+1)$.

g) Let (S,G) be a Hecke pair and assume $\operatorname{char} F = 0$. Given $g \in G$, $g \notin S$, show that the following assertions are equivalent:

(i) $S \cup SgS$ is a subgroup of G .

(ii) $(SgS)^2 = d S + (d-1) SgS$, where $d = \operatorname{ind} SgS$.

(iii) $F S + F SgS$ is a subalgebra of $\mathcal{H}(S,G)$.

In this case one has $\widehat{(SgS)} = SgS$.

4. Quadratic Residues and Non-Residues

Let p be an odd prime and S_p the multiplicative group of non-zero squares in $\mathbb{Z}/p\mathbb{Z}$. Then S_p acts on the additive group $N_p = \mathbb{Z}/p\mathbb{Z}$ by multiplication. Let G_p denote the semi-direct product of N_p with S_p and $\mathcal{H}_p = \mathcal{H}(S_p, G_p)$ the associated Hecke algebra. According to I(7.4) there are exactly 3 double cosets, namely $\{0\} := S_p$, QR_p with the quadratic residues $\bmod p$ as representatives and QN_p with the quadratic non-residues $\bmod p$ as representatives of the right cosets. Let $(\frac{a}{p})$ denote the Legendre symbol. Then $\frac{1}{4}\left(p - (\frac{-1}{p})\right)$ is always an integer. Moreover let F_p denote a splitting field of the polynomial $X^2 + X - \left(\frac{-1}{p}\right)\frac{1}{4}\left(p - (\frac{-1}{p})\right) \in F[X]$. Then the structure of $\mathcal{H}_p$ is described in

(4.1) Lemma. a) *If* $p \equiv 1 \bmod 4$ *the products in* $\mathcal{H}_p$ *are given by*

$$QR_p^{\,2} = \tfrac{p-1}{2}\{0\} + \tfrac{p-5}{4} QR_p + \tfrac{p-1}{4} QN_p , \quad QN_p^{\,2} = \tfrac{p-1}{2}\{0\} + \tfrac{p-1}{4} QR_p + \tfrac{p-5}{4} QN_p ,$$

$$QR_p \cdot QN_p = \tfrac{p-1}{4} QR_p + \tfrac{p-1}{4} QN_p .$$

b) *If* $p \equiv 3 \bmod 4$ *the products in* $\mathcal{H}_p$ *are given by*

$$QR_p^{\,2} = \tfrac{p-3}{4} QR_p + \tfrac{p+1}{4} QN_p , \quad QN_p^{\,2} = \tfrac{p+1}{4} QR_p + \tfrac{p-3}{4} QN_p ,$$

$$QR_p \cdot QN_p = \tfrac{p-1}{2}\{0\} + \tfrac{p-3}{4} QR_p + \tfrac{p-3}{4} QN_p .$$

c) *If* char $F \neq p$ *then* $\mathcal{H}_p$ *is semi-simple and*

$$\mathcal{H}_p \cong \begin{cases} F \times F \times F , & \text{if } \bar{F}_p = F , \\ F \times \bar{F}_p , & \text{if } \bar{F}_p \neq F . \end{cases}$$

d) *If* char $F = p$ *one has*

$$\operatorname{rad} \mathcal{H}_p = F T_0 + F\left(QR_p - QN_p\right) .$$

Proof. Let $p \equiv 1 \mod 4$, i.e., $\left(\frac{-1}{p}\right) = 1$. Since -1 is quadratic residue mod p , we have $\widehat{QR}_p = QR_p$. There are exactly $p-1$ solutions mod p of the congruence

(1) $$a^2 + b^2 \equiv 1 \mod p .$$

This is well-known and can for instance be shown as an application of [54] , 2.3.10 . Omitting the trivial solutions $(\pm 1,0)$, $(0,\pm 1)$, we conclude that the coefficient of QR_p in ${QR_p}^2$ equals $\frac{p-5}{4}$. In view of

$$\operatorname{ind} QR_p = \operatorname{ind} QN_p = \frac{p-1}{2} ,$$

now a) follows from (3.3) .

If char $F \nmid \frac{p-1}{4}$ we consider the left multiplication by QR_p on $\mathcal{H}_p$ with respect to the basis $\{0\}, QR_p, QN_p$ and obtain

$$\mathcal{H}_p = F[QR_p] \cong F[A] , \qquad A = \begin{pmatrix} 0 & \frac{p-1}{2} & 0 \\ 1 & \frac{p-5}{4} & \frac{p-1}{4} \\ 0 & \frac{p-1}{4} & \frac{p-1}{4} \end{pmatrix} .$$

The characteristic polynomial of A equals

$$\left(X - \tfrac{p-1}{2}\right)\left(X^2 + X - \tfrac{p-1}{4}\right) .$$

The roots are mutually distinct, whenever char $F \nmid p$. If char $F = p$ one checks d) by the use of a) . If char $F \mid \frac{p-1}{4}$ then T_0 , $-QR_p$, $-QN_p$ is a complete system of orthogonal idempotents.

Now let $p \equiv 3 \mod 4$, i.e., $\left(\frac{-1}{p}\right) = -1$. In this case we have $\widehat{QR}_p = QN_p$ and the assertion follows from (3.2) in view of $r = p$. □

As an amusing consequence we can conclude from (3.2) resp. (4.1) that the number of solutions mod p of the congruence

$$a^2 + b^2 \equiv t \mod p$$

for $t \not\equiv 0 \mod p$ and $p \equiv 3 \mod 4$ equals $p+1$. For $p \equiv 1 \mod 4$ however we have to know the number of solutions in (1) in order to conclude that the congruence

$$a^2 + b^2 \equiv t \mod p$$

for an arbitrary $t \not\equiv 0 \mod p$ has $p-1$ solutions.

(4.2) Exercises. a) Let p be an odd prime and consider the Hecke algebra $\mathcal{H}_p$ over $\mathbb{Z}$. Show that there exists $T \in \mathcal{H}_p$ satisfying $\mathcal{H}_p = \mathbb{Z}[T]$ iff $p = 3$ or $p = 5$.
b) Describe the corresponding Hecke algebra $\mathcal{H}_2$.
c) Derive the analogous results for an arbitrary finite field $GF(q)$ in place of $\mathbb{Z}/p\mathbb{Z}$ (Hint: [54], 2.3.9).

5. The Canonical Involution on the Hecke Algebra

In this section we study the effect of the canonical involution $\hat{}$ described in I(7.3). We generalize the results in [54], VIII, § 13, for group algebras to Hecke algebras.

Recall the definitions of ordered fields and real closed fields from [54], III, § 1, 2. A quadratic form q on a vector space V is called *isotropic*, if $0 \neq x \in V$ exists with $q(x) = 0$, and *anisotropic* otherwise.

Then (1.6) and I(8.1) lead to

(5.1) Proposition. *Let* S *be a subgroup of a finite group* G *with double cosets* D_i and $d_i = \operatorname{ind} D_i$, $i = 1,\ldots,n$.
a) *Assume that* $\operatorname{char} F = 0$ *and that the quadratic form given by* $\operatorname{diag}(d_1,\ldots,d_n)$ *is anisotropic over* F. *Then every simple factor of the Hecke algebra* $\mathcal{H}(S,G)$ *over* F *is invariant under the canonical involution* $\hat{}$.
b) *Assume that* F *is an ordered field. Then the canonical bilinear form* τ *on* $\mathcal{H}(S,G)$ *is positive definite with respect to this ordering.*

Proof. a) The assumption and I(8.1) imply that τ is anisotropic. $\mathcal{H}(S,G)$ is semi-simple due to (1.6). Let $\mathcal{H}(S,G) = \mathcal{A}_1 \oplus \ldots \oplus \mathcal{A}_m$ be the decomposition into simple components. Hence we have

$$\mathcal{H}(S,G) = \hat{\mathcal{A}}_1 \oplus \ldots \oplus \hat{\mathcal{A}}_m$$

with simple subalgebras $\hat{\mathcal{A}}_i$, too. Since this decomposition is unique up to the order, we have

$$\hat{\mathcal{A}}_i = \mathcal{A}_j \quad \text{for some } j.$$

If $i \neq j$ one has

$$\tau(T,T) = \operatorname{tr}\left(\hat{T}\cdot T\right) = 0 \quad \text{for all } T \in \mathcal{A}_i.$$

This contradicts the fact that τ is anisotropic.
b) The matrix of τ with respect to the basis $D_1,\ldots,D_n$ is given by $\operatorname{diag}(d_1,\ldots,d_n)$ and positive definite. □

The result (5.1) is trivial, whenever the characteristic of F is positive. The quadratic form given by $\mathrm{diag}(d_1,\dots,d_n)$ is always isotropic over F, whenever $n \geq 3$ (cf. [54], 2.3.8). But in the case of $\dim \mathcal{H}(S,G) \leq 2$ the canonical involution $\hat{}$ is the identity.

If $\mathcal{A}, \mathcal{B}$ are two F-algebras with involutions $\bar{}$ resp. * we call an algebra isomorphism $\varphi : \mathcal{A} \to \mathcal{B}$ an *isomorphism of the algebras with involution* if

$$\varphi(\bar{a}) = \varphi(a)^* \quad \text{for all } a \in \mathcal{A} .$$

In this case we write $(\mathcal{A},^-) \cong (\mathcal{B},^*)$.

Given a real closed field F denote its algebraic closure by $F_{\mathbb{C}} = F[\sqrt{-1}]$ and the division algebra of Hamiltonian quaternions over F by $F_{\mathbb{H}}$.

Recall that the prime denotes the transpose of a matrix. We generalize the result in [54], 8.13.3, to

(5.2) Theorem. *Let S be a subgroup of a finite group G and F a real closed field. Then every simple factor $\mathcal{A}$ of the Hecke algebra $\mathcal{H}(S,G)$ is invariant under the canonical involution $\hat{}$. Moreover there exists an isomorphism*

$$(\mathcal{A},\hat{}) \cong (\mathrm{Mat}(n;K), \bar{}\,') ,$$

where $K = F$, $F_{\mathbb{C}}$ or $F_{\mathbb{H}}$ and $\bar{}$ is the identity for $K = F$, complex conjugation for $K = F_{\mathbb{C}}$, and the canonical involution for $K = F_{\mathbb{H}}$.

Proof. $\mathcal{A}$ is invariant under $\hat{}$ according to (5.1). We choose an isomorphism

$$\varphi : \mathcal{A} \to \mathrm{Mat}(n;K) ,$$

where we only have 3 possibilities for K, namely $K = F$, $F_{\mathbb{C}}$, or $F_{\mathbb{H}}$ due to [54], 8.6.4. Let * denote the induced involution on $\mathrm{Mat}(n;K)$, i.e.,

$$X^* = \varphi\left(\widehat{\varphi^{-1}(X)}\right) \quad \text{for } X \in \mathrm{Mat}(n;K) .$$

We define a bilinear form σ on $\mathrm{Mat}(n;K)$ via

$$\text{(1)} \qquad \sigma(X,Y) = \mathrm{tr}\left(\varphi^{-1}\left(X^* \cdot Y\right)\right) = \tau\left(\varphi^{-1}(X), \varphi^{-1}(Y)\right) .$$

By construction φ is an isometry between $(\mathcal{A}, \tau)$ and $(\mathrm{Mat}(n;K), \sigma)$. τ is positive definite on $\mathcal{A}$ according to (5.1) and thus σ on $\mathrm{Mat}(n;K)$.

In view of [54], 8.7.4, there exists $H \in GL(n;K)$ such that

$$X^* = H^{-1}\bar{X}'H \quad \text{for all } X \in \mathrm{Mat}(n;K) ,$$

where $\bar{}$ is an F-involution on K and $\bar{H}' = \pm H$. We show that H is (positive or negative) definite. This is done case-by-case for K.

(i) Let $K = F$. Then $\bar{}$ is the identity. If H fails to be a symmetric definite matrix, we can find $0 \neq x \in F^n$ with $x'Hx = 0$ and therefore $0 \neq X \in \mathrm{Mat}(n;F)$, which satisfies $X'HX = 0$. This leads to

$$X^* X = H^{-1} X' H X = 0$$

and contradicts the fact that σ is positive definite due to (1).
(ii) Let $K = F_{\mathbb{C}}$. If $\bar{\ }$ were the identity, we would obtain

$$\sigma\left(\sqrt{-1}\, I, \sqrt{-1}\, I\right) = -\operatorname{tr}\left(\varphi^{-1}(I)\right) < 0$$

as a contradiction. Hence $\bar{\ }$ is complex conjugation. We may suppose that H is Hermitian and can then proceed as in (i) .
(iii) If $K = F_{\mathbb{H}}$ we can assume that $\bar{\ }$ is the canonical involution. H is not skew Hermitian, since this implies that σ is isotropic. For Hermitian H proceed as in (i) .

Now we may assume that $H = \bar{H}'$ is a positive definite matrix in all the 3 cases, because we may replace H by $-H$. Then $U \in GL(n;K)$ exists satisfying $H = \bar{U}'U$. Now a straightforward calculation shows that

$$\psi : \mathcal{A} \longrightarrow \mathrm{Mat}(n;K) \ ,$$
$$a \longmapsto U\varphi(a)U^{-1} ,$$

becomes an algebra isomorphism, which satifies

$$\overline{\psi(a)}' = \psi(\hat{a}) \ .$$ □

Just as for group algebras (5.2) means that the Hecke algebra over $\mathbb{R}$ contains as many information as the Hecke algebra over $\mathbb{R}$ with involution.

(5.3) Corollary. *Let* S *be a subgroup of a finite group* G *and* F *a real closed field. If the canonical involution* $\hat{\ }$ *is the identity, the Hecke algebra* $\mathcal{H}(S,G)$ *over* F *is commutative and isomorphic to the direct sum of* n *copies of* F , *where* n *is equal to* $\dim \mathcal{H}(S,G)$.

Proof. Due to (5.2) the canonical involution on each simple factor is the identity. Thus each simple factor is isomorphic to F . □

This leads to a classification of three- and four-dimensional Hecke algebras over real closed fields.

(5.4) Corollary. *Let* S *be a subgroup of a finite group* G *and* F *a real closed field.*
a) *If* $\dim \mathcal{H}(S,G) = 3$ *one has*

$$\mathcal{H}(S,G) \cong \begin{cases} F \times F \times F & \text{if } \hat{\ } \text{ is the identity,} \\ F \times F_{\mathbb{C}} & \text{if } \hat{\ } \text{ is not the identity.} \end{cases}$$

b) *If* $\dim \mathcal{H}(S,G) = 4$ *one has*

$$\mathcal{H}(S,G) \cong \begin{cases} F \times F \times F \times F & \text{if } \hat{\ } \text{ is the identity,} \\ F \times F \times F_{\mathbb{C}} & \text{if } \hat{\ } \text{ is not the identity.} \end{cases}$$

Proof. Due to (5.3) let $\hat{}$ be different from the identity. Then a) is clear. In b) we observe that $\mathcal{H}(S,G)$ must be isomorphic to

$$F \times F \times F \times F \quad , \quad \mathrm{Mat}(2;F) \ , \quad F \times F \times F_{\mathbb{C}} \ , \quad F_{\mathbb{C}} \times F_{\mathbb{C}} \ , \quad F_{\mathbb{H}} \ .$$

$\mathrm{Mat}(2;F)$ cannot occur, since $\mathcal{H}(S,G)$ is commutative due to (3.4) . The other cases are not possible in view of

$$\dim \{ T \in \mathcal{H}(S,G) \mid \hat{T} = T \} \ = \ 3$$

such that

$$\mathcal{H}(S,G) \ \cong \ F \times F \times F_{\mathbb{C}}$$

□

Explicit examples of the cases in a) were given in section 4 . The cases in b) already occur as group algebras of $\mathbb{Z}/2\mathbb{Z} \times \mathbb{Z}/2\mathbb{Z}$ resp. $\mathbb{Z}/4\mathbb{Z}$. If one compares (5.4) with (3.2) , one recognizes $\bar{F} = F_{\mathbb{C}}$ for real closed F .

We can also attach another bilinear form. Given $T \in \mathcal{H}(S,G)$ let $\ell(T)$ denote the left multiplication by T on $\mathcal{H}(S,G)$. Then

$$\sigma(T_1,T_2) \ := \ \mathrm{trace}\ \ell\big(\hat{T}_1 \cdot T_2\big)$$

is a bilinear form. If S is normal in G , then σ and τ coincide up to the factor $[G:S]$. But in general they do not coincide.

(5.5) Corollary. *Let* S *be a subgroup of a finite group* G .

a) *The bilinear form* σ *is symmetric.*

b) *Assume that* F *is an ordered field. Then* σ *is positive definite with respect to this ordering. Moreover the restriction of* σ *to any simple factor of* $\mathcal{H}(S,G)$ *is positive definite.*

Proof. We may assume that F is real closed. Consider a simple component $\mathcal{A}$ of $\mathcal{H}(S,G)$, which is invariant under $\hat{}$ due to (5.1) . Given $T_1,T_2 \in \mathcal{A}$ we have

$$\sigma(T_1,T_2) = \mathrm{trace}\ \ell\big(\hat{T}_1 \cdot T_2\big)\Big|_{\mathcal{A}} \ ,$$

because of the decomposition as a direct sum. Hence it suffices to prove that this restriction is a symmetric positive definite bilinear form. According to (5.2) we choose an isomorphism of the algebras with involution

$$\varphi : \mathcal{A} \longrightarrow \mathrm{Mat}(n;K) \quad \text{satisfying} \quad \varphi\big(\hat{T}\big) = \overline{\varphi(T)}' \ .$$

This isomorphism yields

$$\sigma(T_1,T_2) = \mathrm{trace}\ \ell\big(\hat{T}_1 \cdot T_2\big)\Big|_{\mathcal{A}} = \mathrm{trace}\ \ell\Big(\overline{\varphi(T_1)}'\varphi(T_2)\Big)\Big|_{\mathrm{Mat}(n;K)} \ .$$

Note that the right hand side denotes the trace of the left multiplication on $Mat(n;K)$ as an F-linear map. Hence we get

$$\sigma(T_1,T_2) = n\, T_{K/F}\left(\text{trace}\left(\overline{\varphi(T_1)}'\varphi(T_2)\right)\right),$$

where $T_{K/F}$ is the trace of the skew-field extension K/F. This leads to the symmetry of σ. Finally σ is positive definite, because $\overline{\varphi(T)}'\varphi(T)$ is a positive semi-definite matrix and therefore has positive trace for $T \neq 0$. □

If one is interested in quadratic forms, the group algebras of groups, where each element has the order 1 or 2, play a decisive role the description of the Witt ring (cf. [54], II, §9). But for Hecke algebras we do not obtain anything new.

(5.6) Lemma. *Let* S *be a subgroup of a finite group* G *and* F *a field with* $\text{char } F = 0$. *Assume that* $(SgS)^2$ *is a multiple of* S *for each* $g \in G$. *Then* S *is normal in* G. *The factor group* G/S *contains only elements of order* 1 *or* 2 *and is abelian.*

Proof. Assume $F = \mathbb{Q}$ without restriction. Denote the double cosets by $D_1,\ldots,D_n$ and set $d_i = \text{ind } D_i$, $i = 1,\ldots,n$. Then I(4.5) and the assumption imply

$$(2) \qquad D_i^2 = d_i^2 S, \quad i = 1,\ldots,n.$$

Thus I(8.1) leads to $\hat{D}_i = D_i$ for $i = 1,\ldots,n$. We apply I(8.1) once more and obtain

$$r = tr\left(T_0^2\right) = \tau\left(T_0,T_0\right) = d_1^2 + \ldots + d_n^2 .$$

But $r = d_1 + \ldots + d_n$ yields $d_i = 1$ for $i = 1,\ldots,n$. Thus S is normal in G. We conclude from (2) that the square of each element in G/S equals S. Thus G/S is also abelian. □

In the case of positive characteristic the result does not come true any longer.

(5.7) Exercises. a) Find an example of a Hecke algebra $\mathcal{H}(S,G)$ and a field F, where the canonical involution permutes two simple components.

b) Find an example of a three-dimensional Hecke algebra over a field F, where the canonical involution $\hat{}$ is the identity, but $\mathcal{H}(S,G)$ fails to be isomorphic to $F \times F \times F$.

c) Let S be a subgroup of a finite group G and $F = \mathbb{R}$. Consider the symmetric bilinear form

$$\rho\left(T_1,T_2\right) := tr\left(T_1 \cdot T_2\right)$$

on $\mathcal{H}(S,G)$ (cf. I(8.3)). Show that ρ is positive definite if and only if the canonical involution $\hat{}$ is the identity. In the latter case $\rho = \tau$ follows.

d) Give an example of a Hecke algebra $\mathcal{H}(S,G)$ over a field F, where the square of each double coset is a multiple of S, but S fails to be normal in G.

e) Let (S,G) be a Hecke pair with $\dim \mathcal{H}(S,G) = 5$. Show that

$$n := \dim \left\{ T \in \mathcal{H}(S,G) \mid \hat{T} = T \right\} \geq 3 .$$

If the ground field F is real closed, show that

$$\mathcal{H}(S,G) \cong \begin{cases} F \times F \times F \times F \times F & \text{if } n = 5 , \\ F \times F \times F \times F_{\mathbb{C}} & \text{if } n = 4 , \\ F \times F_{\mathbb{C}} \times F_{\mathbb{C}} & \text{if } n = 3 . \end{cases}$$

f) Let S be a subgroup of a finite group G and set $n = \dim \mathcal{H}(S,G)$. Assume that $S \cup SgS$ is a subgroup of G for each $g \in G$. Show that the Hecke algebra $\mathcal{H}(S,G)$ is commutative. If F is real closed, show that $\mathcal{H}(S,G)$ is isomorphic to n copies of F .

Remarks

The version of Maschke's Theorem in (1.6) was proved in [36] . The weaker version (2.1) seems to be due to N. Iwahori [24] , confer also N. Bourbaki [9] . A more general version of Maschke's Theorem referring to double cosets with respect to two commensurable subgroups of a group G is dealt with in [40] . The results of section 2 are partly contained in the book of C.W. Curtis and I. Reiner [16] and are consequences of the results on group algebras as for instance described by D.S. Passman [51] . The results in section 5 follow from W. Scharlau's procedure [54] for group algebras.

Chapter III

Examples of Hecke Algebras for Finite Groups

In this chapter we will deal with several examples of Hecke algebras for finite groups. The groups naturally arise in Algebra or Number Theory. The aim is to describe the arising Hecke algebra and its structure as explicitly as possible.

We start with the dihedral group. Then we deal with the integral orthogonal group, which contains the symmetric group as a canonical subgroup. Then we generalize our first example described in section I.5 . We treat the symmetric group and two types of subgroups.

Finally we deal with the general linear group over a finite field. Here the Hecke algebras for two important subgroups are described explicitly. The first one is a Borel subgroup or minimal parabolic subgroup. The second one is a maximal parabolic subgroup. As an application of the results of C.W. Curtis, N. Iwahori, and R. Kilmoyer [15] we can show that the Hecke algebra associated with an arbitrary parabolic subgroup $\mathcal{P}$ of the general linear group is commutative if and only if $\mathcal{P}$ is maximal.

Again we suppose throughout the whole chapter that the ground ring

$$R = F \quad \text{is a field.}$$

1. The Dihedral Group

Let G_n denote the dihedral group of order $2n$. Then G_n is generated by two elements a and b, where a has the order n and b the order 2 and the characteristic equation is given by

$$bab = a^{-1} .$$

Thus G_n equals the semi-direct product of the groups generated by a and b . Choose the subgroup $S = \{e,b\}$, which has the index n in G_n . Consider the double cosets

$$A_i = Sa^iS \quad , \quad i \in \mathbb{Z} .$$

Clearly one has

$$A_i = A_{-i} = A_{n+i} .$$

Obviously the following hold

$$A_0 = S \quad , \qquad \text{ind } A_0 = 1 ,$$

$$A_i = Sa^i \cup Sa^{-i} \quad , \qquad \text{ind } A_i = 2 \quad , \quad \text{for } 0 < i < \tfrac{n}{2} ,$$

$$A_{n/2} = Sa^{n/2} \quad , \qquad \text{ind } A_{n/2} = 1 \quad , \quad \text{if } n \text{ is even} .$$

We observe that $A_i = \hat{A}_i = A_{-i}$ implies that the Hecke algebra $\mathcal{H}(S,G_n)$ is commutative according to I(7.3) , where we could also apply I(7.4) . The right coset decompositions immediately yield the following products

(1) $$A_i \cdot A_j = \left(\frac{4}{\# A_{i+j}}\right) A_{i+j} + \left(\frac{4}{\# A_{i-j}}\right) A_{i-j} \quad \text{for } 0 < i,j < \tfrac{n}{2} ,$$

(2) $$A_i \cdot A_{n/2} = A_{(n/2)-i} \qquad \text{for all } i , \text{ if } n \text{ is even.}$$

Here $(4/\# A_k)$ is always an integer, because the order of a double coset divides 4 . If $[x]$ denotes the largest integer less than or equal to x , we obtain the following result.

(1.1) Proposition. a) $\mathcal{H}(S,G_n)$ *is a commutative algebra of dimension* $[\frac{n}{2}] + 1$.
b) $\mathcal{H}(S,G_n)$ *consists of all polynomials over* F *in* A_1 *, i.e.,*

$$\mathcal{H}(S,G_n) = F[A_1] ,$$

except for char $F = 2$ *and even* $n > 2$.
c) $\mathcal{H}(S,G_n)$ *is semi-simple if and only if* char $F \nmid n$.

Proof. a) $$G_n = \bigcup_{0 \le i \le \frac{n}{2}} A_i$$

holds as a disjoint union.
b) Apply (1) . If $n = 2k > 2$ is even, we obtain

$$F[A_1] = \left\{ t_0 A_0 + t_1 A_1 + \ldots + t_{k-1} A_{k-1} + 2t_k A_k \mid t_i \in F , \ 0 \le i \le k \right\} .$$

c) Due to II(1.3) and II(1.6) the assertion is clear except for

$$n = 2k + 1 \text{ odd and char } F = 2 .$$

In this case let $T = t_0 A_0 + \ldots + t_k A_k \in \mathcal{H}(S,G_n)$ be a nilpotent element. Due to (1) and char $F = 2$ we get

$$T^2 = t_0^2 A_0^2 + t_1^2 A_1^2 + \ldots + t_k^2 A_k^2 = \sum_{i=0}^{k} t_i^2 A_{2i} .$$

In view of $A_i = A_{n-i}$ and $n = 2k + 1$ we have

$$\{A_0, A_2, \ldots, A_{2k}\} = \{A_0, A_1, \ldots, A_k\} .$$

Thus $T = 0$ is the only nilpotent element in $\mathcal{H}(S,G_n)$ and $\mathcal{H}(S,G_n)$ over F proves to be semi-simple. □

For small values of n one can describe the structure explicitly. Moreover it follows from II(5.3) that $\mathcal{H}(S,G_n)$ over $\mathbb{R}$ is always isomorphic to $[\frac{n}{2}]+1$ copies of $\mathbb{R}$.

(1.2) Exercises. a) Show that

(i) $\mathcal{H}(S,G_2) \cong F \times F$, if $\operatorname{char} F \neq 2$,

(ii) $\mathcal{H}(S,G_3) \cong F \times F$, if $\operatorname{char} F \neq 3$,

(iii) $\mathcal{H}(S,G_4) \cong F \times F \times F$, if $\operatorname{char} F \neq 2$,

(iv) $$\mathcal{H}(S,G_5) \cong \begin{cases} F \times F \times F, & \text{if } 5 \text{ is a square in } F, \quad \operatorname{char} F \nmid 10, \\ F \times F[\sqrt{5}], & \text{if } 5 \text{ is not a square in } F, \operatorname{char} F \nmid 10. \end{cases}$$

b) Describe the case of $\operatorname{char} F = 2$ in (iv). Determine the radical in those cases of a), where the Hecke algebra fails to be semi-simple.

c) Let $n = 2k$ be even. Show that the map

$$\mathcal{H}(S,G_n) \to \mathcal{H}(S,G_n),$$
$$\sum_{i=0}^{k} t_i A_i \mapsto \sum_{i=0}^{k} (-1)^i t_i A_i,$$

is an involution of the Hecke algebra $\mathcal{H}(S,G_n)$.

d) Let $n = 2k$ be even. Verify that $\bar{S} = S \cup A_k$ is a subgroup of G_n. Show that the Hecke algebra $\mathcal{H}(\bar{S},G_n)$ is isomorphic to the subalgebra

$$\{T \in \mathcal{H}(S,G_n) \mid T \cdot A_k = T\}$$

of $\mathcal{H}(S,G_n)$, whenever $\operatorname{char} F \neq 2$.

2. The Integral Orthogonal Group

We deal with the case $G = \mathcal{O}(n;\mathbb{Z})$, the integral orthogonal group of degree n. Therefore let $I = I^{(n)}$ always denote the $n \times n$ identity matrix and U' the transpose of a matrix U. Set

$$\mathcal{O}(n;\mathbb{Z}) = \{U \in \operatorname{Mat}(n;\mathbb{Z}) \mid U'U = UU' = I\}.$$

If $u_1,\dots,u_n$ are the column vectors of a matrix $U \in \mathcal{O}(n;\mathbb{Z})$ then $U'U = I$ implies $u_j'u_j = 1$. Hence u_j possesses a single ±1 and 0's otherwise. Since U is invertible, we have once ±1 and 0's otherwise in each row and column. If $e_1,\dots,e_n$ is the elementary basis of $\mathbb{Z}^n$, we can describe the elements of $\mathcal{O}(n;\mathbb{Z})$ in the form

(1) $$(\varepsilon_1 e_{\pi(1)},\dots,\varepsilon_n e_{\pi(n)}), \quad \pi \in S_n, \quad \varepsilon_i = \pm1, \quad i = 1,\dots,n.$$

Thus we calculate

$$\# \mathcal{O}(n;\mathbb{Z}) = 2^n n! \ .$$

There exists a subgroup of all permutation matrices. Given $\pi \in S_n$ set

$$P_\pi = \left(e_{\pi(1)}, \dots, e_{\pi(n)}\right) = \sum_{j=1}^{n} I_{\pi(j),j} \in \mathcal{O}(n;\mathbb{Z}) \ ,$$

where the $I_{j,k}$ are the canonical matrix units in $\mathrm{Mat}(n;\mathbb{Z})$ with a single 1 as its (j,k)-entry and 0 's otherwise. A straightforward calculation gives

$$P_{\pi\circ\sigma} = P_\pi P_\sigma \ , \quad P_\pi^{-1} = P_{\pi^{-1}} = P'_\pi \ , \quad \det P_\pi = \operatorname{sign} \pi$$

for all $\pi,\sigma \in S_n$. We denote the subgroup of permutation matrices also by S_n. Moreover (1) yields that $\mathcal{O}(n;\mathbb{Z})$ equals the semi-direct product of S_n and $\{\pm 1\}^n$.

Claim 1. *If* $n > 1$ *then* S_n *fails to be normal in* $\mathcal{O}(n;\mathbb{Z})$.

Proof. $\begin{pmatrix} -1 & 0 \\ 0 & 1 \end{pmatrix} \begin{pmatrix} 0 & 1 \\ 1 & 0 \end{pmatrix} \begin{pmatrix} -1 & 0 \\ 0 & 1 \end{pmatrix} = \begin{pmatrix} 0 & -1 \\ -1 & 0 \end{pmatrix} \notin S_2$.

Use a similar calculation for $n > 2$. □

Observe that multiplication by an element of S_n from the left hand side has the effect of permuting the rows. Hence a set of representatives of the 2^n right cosets relative to S_n in $\mathcal{O}(n;\mathbb{Z})$ is given by

$$\operatorname{diag}(d_1, \dots, d_n) \ , \quad d_i = \pm 1 \ , \quad i = 1, \dots, n \ .$$

Since S_n acts from the right hand side by permuting the columns, we get

Claim 2. *There are* $n+1$ *double cosets relative to* S_n *in* $\mathcal{O}(n;\mathbb{Z})$, *namely*

$$D_k = S_n \operatorname{diag}(\underbrace{-1, \dots, -1}_{k}, \underbrace{1, \dots, 1}_{n-k}) S_n \ , \quad k = 0, \dots, n \ .$$

Hence k *denotes the number of* -1 *'s in the matrices of a double coset.* D_0 *is equal to* S_n *and one has*

$$\operatorname{ind} D_k = \binom{n}{k} \ .$$

Due to Claim 2 one has $\hat{D}_k = D_k$. Thus the Hecke algebra $\mathcal{H}(S_n, \mathcal{O}(n;\mathbb{Z}))$ is commutative according to I(7.3) resp. I(7.4) .

Each double coset D_k , $0 \le k \le n$, decomposes into $\binom{n}{k}$ right cosets

(2) $\qquad S_n \operatorname{diag}(d_1, \dots, d_n),\ 1 \le i_1 < \dots < i_k \le n ,\ d_{i_j} = -1 ,\ d_j = 1$ otherwise.

In particular D_n consists of a single left and right coset. Now a combination of Claim 2 and I(4.6) yields

(3) $\qquad D_k \cdot D_n = D_{n-k} \ , \quad 0 \le k \le n \ .$

Due to (3) the following result describes the product in $\mathcal{H}(S_n, O(n;\mathbb{Z}))$ completely.

Claim 3. *Given* $0 \le k \le \ell \le \frac{n}{2}$ *one has*

$$D_\ell \cdot D_k = \sum_{j=0}^{k} \binom{\ell-k+2j}{j} \binom{n+k-\ell-2j}{k-j} D_{\ell-k+2j} \,.$$

Proof. Set

$$D_\ell \cdot D_k = \sum_{i=0}^{n} t_i D_i \,.$$

Consider the representatives of the right coset in the form (2) and of the double cosets in the form of Claim 2. In view of the definition of the product t_i equals the number of matrices

$$\operatorname{diag}(d_1, \dots, d_n) \text{ such that } d_\nu = \pm 1 \text{ and } \#\{\nu \mid 1 \le \nu \le n,\ d_\nu = -1\} = k$$
$$\text{as well as } \#\{\nu \mid 1 \le \nu \le i,\ -d_\nu = -1\} + \#\{\nu \mid i < \nu \le n,\ d_\nu = -1\} = \ell.$$

Suppose that the number of -1's among $d_1, \dots, d_i$ equals j, $0 \le j \le k$. Then there must be $(k-j)$-times -1 among $d_{i+1}, \dots, d_n$. Hence there are $\binom{i}{j}\binom{n-i}{k-j}$-possibilities for $\operatorname{diag}(d_1, \dots, d_n)$. Counting the number of -1's among $-d_1, \dots, -d_i, d_{i+1}, \dots, d_n$ we get

$$(i-j) + (k-j) = \ell \,, \quad \text{hence } i = \ell-k+2j \,. \qquad \square$$

In particular the formula gives the square of a double coset.

$$(4) \qquad D_k^2 = D_{n-k}^2 = \sum_{j=0}^{\min(k,n-k)} \binom{2j}{j} \binom{n-2j}{k-j} D_{2j} \,, \quad 0 \le k \le n \,.$$

On the other hand we combine (3) and Claim 3 in order to get

$$(5) \qquad D_\ell \cdot D_1 = (n+1-\ell) D_{\ell-1} + (\ell+1) D_{\ell+1} \qquad \text{for } 0 < \ell < n \,.$$

(2.1) Theorem. a) *The Hecke algebra* $\mathcal{H}(S_n, O(n;\mathbb{Z}))$ *is commutative and has the dimension* $n+1$.

b) *Assume that* $\operatorname{char} F \nmid 2 \cdot n!$. *Then*

$$\mathcal{H}(S_n, O(n;\mathbb{Z})) = F[D_1]$$

holds. $\mathcal{H}(S_n, O(n;\mathbb{Z}))$ *over* F *is semi-simple and isomorphic to the direct sum of* $n+1$ *copies of* F.

Proof. a) Use Claim 2.

b) $$\mathcal{H}(S_n, O(n;\mathbb{Z})) = F[D_1]$$

follows from (5). The Hecke algebra is semi-simple due to II(2.1). In order to determine the precise structure, we consider the left multiplication by D_1 on $\mathcal{H}(S_n, O(n;\mathbb{Z}))$. The matrix of the left multiplication with respect to the basis $D_0, \dots, D_n$ is given by

$$\begin{pmatrix} 0 & n & & & & 0 \\ 1 & 0 & n-1 & & & \\ & 2 & 0 & \ddots & & \\ & & 3 & \ddots & \ddots & \\ & & & \ddots & \ddots & 2 \\ & & & & \ddots & 0 & 1 \\ 0 & & & & & n & 0 \end{pmatrix}$$

according to (5) and (3) . The characteristic polynomial of this matrix was already computed by J.J. Sylvester [61] and A. Cayley [13] and equals

$$\prod_{\substack{-n\le \ell\le n \\ \ell\equiv n \bmod 2}} (X - \ell) .$$

Hence the matrix has $n+1$ mutually different eigenvalues in F . Thus $F[D_1]$ is isomorphic to the direct sum of $n+1$ copies of F . □

Due to II(1.3) the Hecke algebra fails to be semi-simple for char $F = 2$. In this case one can however describe the radical. Because of $\binom{2j}{j} = 2\binom{2j-1}{j}$ for $j \ge 1$, we get

$$T^2 = (\operatorname{ind} T)^2\, D_0 \qquad \text{for char } F = 2$$

from (4) . Thus one has

$$\begin{aligned} \operatorname{rad} \mathcal{H}(S_n, \mathcal{O}(n;\mathbb{Z})) &= \left\{ T \in \mathcal{H}(S_n, \mathcal{O}(n;\mathbb{Z})) \mid \operatorname{ind} T = 0 \right\} \\ &= \sum_{k=1}^{n} F\left(D_k - \binom{n}{k} D_0\right) \end{aligned}$$

as well as

$$\mathcal{H}(S_n,\mathcal{O}(n;\mathbb{Z}))\big/_{\operatorname{rad} \mathcal{H}(S_n,\mathcal{O}(n;\mathbb{Z}))} \cong F .$$

If $2 < \text{char } F = p \le n$ then $F[D_1]$ is a p-dimensional subalgebra of $\mathcal{H}(S_n,\mathcal{O}(n;\mathbb{Z}))$. But the structure of the Hecke algebra is not clear. In this case however it is obvious that $\mathcal{H}(S_n,\mathcal{O}(n;\mathbb{Z}))$ is generated by $D_1, D_p, D_{2p}, \ldots, D_{rp}$, where $r = [\frac{n}{p}]$.

This example is important in the representation theory of finite groups, since $\mathcal{O}(n;\mathbb{Z})$ is the Weyl group of the symplectic group of degree n and S_n is the Weyl group of the canonical maximal parabolic subgroup. But the relation with Number Theory can be described as follows. Given a real positive definite $n \times n$ matrix S , then

$$\operatorname{Aut} S = \{ U \in \operatorname{Mat}(n;\mathbb{Z}) \mid U'SU = S \}$$

proves to be a finite subgroup of the unimodular group

$$GL(n;\mathbb{Z}) = \{ U \in \operatorname{Mat}(n;\mathbb{Z}) \mid \det U = \pm 1 \} .$$

One has $\operatorname{Aut} I = \mathcal{O}(n;\mathbb{Z})$. Consider $e = (1,\ldots,1)' \in \mathbb{Z}^n$ and the positive definite matrix $P = I + ee'$. The group $\operatorname{Aut} P$ is isomorphic to $\{\pm 1\} \times S_{n+1}$. Moreover $\operatorname{Aut} I$ and $\operatorname{Aut} P$

are maximal finite subgroups of $GL(n;\mathbb{Z})$ (cf. M. Newman [49] or M. Koecher [32]). One can check that

(6) $$\text{Aut } I \cap \text{Aut } P = \{\pm P_\pi \mid \pi \in S_n\} .$$

Thus the Hecke algebra $\mathcal{H}(\text{Aut } I \cap \text{Aut } P, \text{Aut } P)$ was described in section I.5 . In the case char $F = 0$ one can use I(6.6) in order to see that $\mathcal{H}(\text{Aut } P \cap \text{Aut } I, \text{Aut } I)$ is isomorphic to the subalgebra

(7) $$\{t_0 D_0 + \ldots + t_n D_n \mid t_i \in F ,\ t_i = t_{n-i} ,\ i = 0,\ldots,n\}$$
$$= \{T \in \mathcal{H}(S_n, \mathcal{O}(n;\mathbb{Z})) \mid T \cdot D_n = T\}$$

of $\mathcal{H}(S_n, \mathcal{O}(n;\mathbb{Z}))$.

(2.2) Exercises. a) Show that $\mathcal{H}(S_3, \mathcal{O}(3;\mathbb{Z}))$ is isomorphic to $F \times F \times F \times F$ even in the case of char $F = 3$.

b) Show that the map

$$\mathcal{H}(S_n, \mathcal{O}(n;\mathbb{Z})) \longrightarrow \mathcal{H}(S_n, \mathcal{O}(n;\mathbb{Z})) ,$$
$$\sum_{i=0}^{n} t_i D_i \mapsto \sum_{i=0}^{n} (-1)^i t_i D_i ,$$

is an involution.

c) Show that the set

$$\mathcal{A} = \{\sum_{0 \le i \le n/2} t_i D_{2i} \mid t_i \in F\}$$

is a subalgebra of $\mathcal{H}(S_n, \mathcal{O}(n;\mathbb{Z}))$. If $n > 1$ and char $F \nmid n!$ demonstrate $\mathcal{A} = F[D_2]$.

d) Show that $\pm D_0, \pm D_2$ are the only units in the Hecke algebra $\mathcal{H}(S_2, \mathcal{O}(2;\mathbb{Z}))$ over $\mathbb{Z}$.

e) Verify (6) and show that the Hecke algebra $\mathcal{H}(\text{Aut } P \cap \text{Aut } I , \text{Aut } I)$ is isomorphic to the subalgebra (7) of $\mathcal{H}(S_n, \mathcal{O}(n;\mathbb{Z}))$, whenever char $F \neq 2$.

3. The Symmetric Group

In this section we consider a more general case than in section I.5 . Therefore let $0 < m < n$ and consider S_m as a subgroup of S_n .

(3.1) Lemma. *Let* $0 < m < n$.

a) *The Hecke algebra* $\mathcal{H}(S_m, S_n)$ *as the dimension*

$$\sum_{\nu=\ell}^{n-m} \nu! \binom{n-m}{\nu}^2 , \quad \ell = \max\{0, n-2m\} .$$

b) $\mathcal{H}(S_m, S_n)$ *contains a subalgebra, which is isomorphic to the group algebra* $F[S_{n-m}]$.

c) *If* $n-m > 1$, *the Hecke algebra* $\mathcal{H}(S_m, S_n)$ *fails to be commutative.*

Proof. a) Let $\mathcal{J}$ be a subset of $\{m+1,\ldots,n\}$ of order ν and let $\Phi : \mathcal{J} \longrightarrow \{m+1,\ldots,n\}$

be a injective mapping. If $\nu \geq \ell$, the set

$$\{\pi \in S_n \mid \pi|_{\mathcal{Q}} = \Phi \ , \ \#(\pi(\{m+1,\dots,n\}) \cap \{m+1,\dots,n\}) = \nu\}$$

forms exactly one double coset relative to S_m in S_n. There are $\binom{n-m}{\nu}$ subsets $\mathcal{Q}$ of order ν. The number of injective mappings of $\mathcal{Q}$ into $\{m+1,\dots,n\}$ equals

$$\frac{(n-m)!}{(n-m-\nu)!} = \nu!\binom{n-m}{\nu} .$$

In analogy with I.5 one sees that these are all the double cosets relative to S_m in S_n.
b) Given $\pi \in S_{n-m}$ define $\psi(\pi) \in S_n$ by

$$\psi(\pi)(\nu) = \nu \quad , \quad 1 \leq \nu \leq m \quad , \quad \psi(\pi)(m+\nu) = m + \pi(\nu) \quad , \quad \nu \geq 1 .$$

Hence the double coset $S_m\psi(\pi)S_m$ consists of a single left and right coset. Thus I(4.6) yields that

$$\left\{ \sum_{\pi \in S_{n-m}} t(\pi)S_m\psi(\pi)S_m \mid t(\pi) \in F \right\}$$

becomes a subalgebra of $\mathcal{H}(S_m,S_n)$, which is isomorphic to the group algebra $F[S_{n-m}]$.
c) If $n-m \geq 2$, we compute

$$(S_m\tau_{n-1,n}S_m)\cdot(S_m\tau_{1,n}S_m) = S_m\tau_{n-1,n}\circ\tau_{1,n}S_m ,$$

$$(S_m\tau_{1,n}S_m)\cdot(S_m\tau_{n-1,n}S_m) = S_m\tau_{1,n}\circ\tau_{n-1,n}S_m$$

in view of I(4.6), since $S_m\tau_{n-1,n}S_m$ consists of a single left and right coset. The two arising double cosets are distinct, since one can for instance compare the values at n or at $n-1$. □

According to the connection with group algebras and II(3.4) one might wonder whether every Hecke algebra, whose dimension is a prime number, is commutative. But the answer is negative, since $\mathcal{H}(S_{n-2},S_n)$, $n \geq 4$, are examples of non-commutative Hecke algebras of dimension 7.

Next let us consider another generalization of I.5. Therefore let $n,m \in \mathbb{N}$ and consider the subgroup

$$S_m \times S_n = \{\pi \in S_{m+n} \mid \pi(\{1,\dots,m\}) = \{1,\dots,m\}\}$$

of S_{m+n}. The subgroups $S_m \times S_n$ and $S_n \times S_m$ are conjugate in S_{m+n}, hence we will always assume $1 \leq m \leq n$. The index of $S_m \times S_n$ in S_{m+n} equals $r = \binom{m+n}{m}$ and all the right cosets relative to $S_m \times S_n$ are given by

(1) $$R_T = \{\pi \in S_{m+n} \mid \pi(T) = \{1,\dots,m\}\} , \text{ where } T \subset \{1,\dots,m+n\} , \#(T) = m .$$

There are exactly $m+1$ double cosets relative to $S_m \times S_n$ in S_{m+n}, namely

(2) $$C_k = (S_m \times S_n)\tau_{1,m+1} \circ \dots \circ \tau_{k,m+k} (S_m \times S_n) , \; k = 0,\dots,m ,$$

where clearly $C_0 = S_m \times S_n$. A right coset R_T of the form (1) satisfies

(3) $$R_T \subset C_k \iff \#(T \cap \{m+1, \dots ,m+n\}) = k .$$

Hence one has

(4) $$\operatorname{ind} C_k = \binom{m}{k}\binom{n}{k} , \quad 0 \le k \le m .$$

We obtain $\hat{C}_k = C_k$ from

$$(\tau_{1,m+1}\circ \dots \circ \tau_{k,m+k})^{-1} = \tau_{1,m+1}\circ \dots \circ \tau_{k,m+k} .$$

Therefore the Hecke algebra $\mathcal{H}(S_m \times S_n, S_{m+n})$ turns out to be commutative due to I(7.3) .

According to (1) , (2) , (3) , and (4) a set of representatives of the right cosets relative to $S_m \times S_n$ in C_1 is given by the transpositions

(5) $$\tau_{i,m+j} , \quad 1 \le i \le m , \; 1 \le j \le n .$$

This is used in order to derive

(3.2) Lemma. *Given* $1 \le m \le n$ *one has the following products in the Hecke algebra* $\mathcal{H}(S_m \times S_n, S_{m+n})$

$$C_k \cdot C_1 = (m+1-k)(n+1-k)\, C_{k-1} + k(m+n-2k)\, C_k + (k+1)^2\, C_{k+1}$$

for $0 < k < m$ *as well as*

$$C_m \cdot C_1 = (n+1-m) C_{m-1} + m(n-m)\, C_m .$$

Proof. Let $0 \le \ell \le m$, $1 \le i \le m$, $1 \le j \le n$. Now one can use (3) in order to verify

$$\tau_{1,m+1}\circ \dots \circ \tau_{\ell,m+\ell}\circ \tau_{i,m+j}^{-1} \in \begin{cases} C_{\ell-1} & \text{if } i \le \ell ,\; j \le \ell , \\ C_\ell & \text{if } i > \ell ,\; j \le \ell , \text{ or} \\ & \text{if } i \le \ell ,\; j > \ell , \\ C_{\ell+1} & \text{if } i > \ell ,\; j > \ell . \end{cases}$$

$C_{\ell-1}$ appears ℓ^2-times, C_ℓ exactly $\ell(m+n-2\ell)$-times and $C_{\ell+1}$ exactly $(m-\ell)(n-\ell)$-times on the right hand side, if $\tau_{i,m+j}$ runs through the set (5) of representatives of the right cosets in C_1 . Using I(4.4 i) we immediately obtain the assertion. □

We describe the structure of the Hecke algebra in

(3.3) Proposition. *Given* $1 \le m \le n$ *the Hecke algebra* $\mathcal{H}(S_m \times S_n, S_{m+n})$ *is a commutative algebra of dimension* $m+1$. *If* $\operatorname{char} F \nmid m!$, *one has*

$$\mathcal{H}(S_m \times S_n, S_{m+n}) = F[C_1] .$$

Proof. Apply the considerations above and (3.2) . □

The case $m = 1$ was already dealt with in section I.5 . Therefore let us have a closer look at $m = 2$, $n \geq 2$. Using (3.2) , (4) , and II(3.3) we get the products

$$(6) \qquad \begin{aligned} C_1^2 &= 2n\,C_0 + n\,C_1 + 4\,C_2 , \qquad C_1 \cdot C_2 = (n-1)\,C_1 + 2(n-2)\,C_2 , \\ C_2^2 &= \binom{n}{2} C_0 + \binom{n-1}{2} C_1 + \binom{n-2}{2} C_2 . \end{aligned}$$

In this particular case we obtain the following structure.

(3.4) Proposition. *Let* $n \geq 2$ *. The Hecke algebra* $\mathcal{H}(S_2 \times S_n, S_{n+2})$ *over* F *is semi-simple if and only if*

$$\operatorname{char} F \nmid \tfrac{1}{2} n(n+1)(n+2) .$$

In this case $\mathcal{H}(S_2 \times S_n, S_{n+2})$ *is isomorphic to* $F \times F \times F$.

Proof. If $\operatorname{char} F \mid r = \frac{1}{2}(n+1)(n+2)$, then $T_0 = C_0 + C_1 + C_2$ is nilpotent due to II(1.2). If $\operatorname{char} F \mid n$ then $C_1 + 2\,C_2$ proves to be nilpotent according to (6) .

Now assume that $\operatorname{char} F \nmid \frac{1}{2} n(n+1)(n+2)$. If $\operatorname{char} F = 2$ we conclude $n \equiv 1 \mod 4$. Then (6) shows that T_0, C_1, C_2 is a complete set of orthogonal idempotents . This yields the isomorphy with $F \times F \times F$.

Therefore let $\operatorname{char} F \nmid n(n+1)(n+2)$. Then we have

$$\mathcal{H}(S_2 \times S_n, S_{n+2}) = F[C_1] \cong F[A] , \quad \text{where } A = \begin{pmatrix} 0 & 2n & 0 \\ 1 & n & n-1 \\ 0 & 4 & 2n-4 \end{pmatrix} ,$$

if we consider the left multiplication by C_1 with respect to the basis C_0, C_1, C_2 . The eigenvalues of A are -2 , $n-2$, and $2n$. Since the eigenvalues are mutually distinct in F , we conclude

$$F[A] \cong F \times F \times F .$$ □

Let in particular $p = n$ be an odd prime and equal to $\operatorname{char} F$. Then we have an example, where the condition (2) in section II.1 is satisfied, but the Hecke algebra fails to be semi-simple.

We can also consider a more general situation, which includes the cases (3.1) and (3.2) . Therefore let $n_1, \dots, n_k \in \mathbb{N}$, set $n = n_1 + \dots + n_k$ as well as $I_1 = \{1, \dots, n_1\}$ and $I_j = \{ n_1 + \dots + n_{j-1} + i \mid 1 \leq i \leq n_j \}$ for $1 < j \leq k$. Consider the subgroup

$$S := S_{n_1} \times \dots \times S_{n_k} := \left\{ \pi \in S_n \mid \pi(I_j) = I_j , \ j = 1, \dots, k \right\}$$

of S_n . Clearly our investigations above refer to the case $k = 2$, where the attached Hecke algebra is commutative due to (3.3) . The case (3.1) is equivalent with $n_1 = m$, $n_2 = \ldots = n_k = 1$, $k = n+1-m$, where the Hecke algebra fails to be commutative for $k > 2$. We can generalize this result.

(3.5) Lemma. *Let* $n_1,\ldots,n_k \in \mathbb{N}$ *and* $n = n_1+\ldots+n_k$ *, where* $k > 2$ *. Then the Hecke algebra* $\mathcal{H}(S,S_n)$ *, where* $S = S_{n_1} \times \ldots \times S_{n_k}$ *, fails to be commutative.*

Proof. Consider $T_1 := S\tau_{1,n_1+1}S$ and $T_2 := S\tau_{1,n_1+n_2+1}S$. In analogy with (5) one can show that a set of representatives of the right cosets relative to S in T_1 resp. T_2 is given by the transpositions

$$\tau_{i,j} \text{ , where } 1 \le i \le n_1 \text{ and } n_1 < j \le n_1+n_2 \text{ resp. } n_1+n_2 < j \le n_1+n_2+n_3 .$$

Thus one can compute the product by application of the definition. Set

$$\rho = \tau_{1,n_1+1} \circ \tau_{1,n_1+n_2+1} \text{ with } \rho^2 = \rho^{-1} , \quad \sigma = \tau_{1,n_1+1} \circ \tau_{2,n_1+n_2+1} .$$

We obtain

$$T_1 \cdot T_2 = \begin{cases} S\rho S + S\sigma S & \text{if } n_1 > 1 , \\ S\rho S & \text{if } n_1 = 1 . \end{cases}$$

The application of the canonical involution I(7.3) and $\hat{T}_i = T_i$ lead to

$$T_2 \cdot T_1 = \widehat{(T_1 \cdot T_2)} = \begin{cases} S\rho^{-1}S + S\sigma S & \text{if } n_1 > 1 , \\ S\rho^{-1}S & \text{if } n_1 = 1 . \end{cases}$$

But the double cosets $S\rho S$ and $S\rho^{-1}S$ are distinct, because

$$\pi(\{1,\ldots,n_1\}) \cap \{n_1+1,\ldots,n_1+n_2\}$$

is empty for $\pi \in S\rho S$ and consists of a single element for $\pi \in S\rho^{-1}S$. □

(3.6) Exercises. a) Determine the radical in the non-semi-simple cases of (3.4) . Using the abbreviation $\mathcal{H} = \mathcal{H}(S_2 \times S_n, S_{n+2})$, show that

(i) $\quad \operatorname{rad} \mathcal{H} = F(C_0 + C_2) + F\,C_1 \quad$ if char $F = 2$, $n \equiv 2 \mod 4$,

(ii) $\quad \operatorname{rad} \mathcal{H} = F(C_1 + 2C_2) \quad$ if char $F = 2$, $n \equiv 0 \mod 4$ or if $2 \ne \operatorname{char} F \mid n$,

(iii) $\quad \operatorname{rad} \mathcal{H} = F T_0 \quad$ if char $F = 2$, $n \equiv 3 \mod 4$ or if $2 \ne \operatorname{char} F \mid (n+1)(n+2)$.

In these cases $\mathcal{H}/\operatorname{rad}\mathcal{H}$ is isomorphic to F resp. $F \times F$.

b) Show that $\pm S$, $\pm C_2$ are the only units in the Hecke algebra $\mathcal{H}(S_2 \times S_2, S_4)$ over $\mathbb{Z}$.

c) Discuss the Hecke algebras $\mathcal{H}(S_3 \times S_3, S_6)$ and $\mathcal{H}(S_3 \times S_4, S_7)$ in analogy with (3.4) .

d) Let the assumptions of (3.2) hold and assume $m \ge 2$. Compute all the products $C_k \cdot C_2$, $0 \le k \le m$.

4. A Borel Subgoup of the General Linear Group over a Finite Field

In this and the following section we consider the general linear group over a finite field. First of all the subgroup in question is a Borel group or minimal parabolic subgroup. The decisive tool is the Bruhat decomposition. The attached Hecke algebra in the more general context of a Chevalley group over a finite field was investigated by N. Iwahori [24] resp. H. Matsumoto [46] .

At first let K be an arbitrary field. Given $n \in \mathbb{N}$ the *Borel subgroup* $\mathcal{B} = \mathcal{B}(n;K)$ consists of all upper triangular matrices in $GL(n;K)$, i.e. ,

$$\mathcal{B} = \{ B = (b_{ij}) \in GL(n;K) \mid b_{ij} = 0 \text{ for } i > j \} .$$

Let I_{jk} denote the canonical matrix units in $Mat(n;K)$. Then $\mathcal{B}$ is obviously generated by the matrices

(1) $\quad \operatorname{diag}(b_1,\dots,b_n) , \quad 0 \neq b_i \in K , \ i = 1,\dots,n ,$

as well as

(2) $\quad I + I_{ij} , \quad 1 \le i < j \le n .$

Due to (1), (2) the effect of multiplication by an element of $\mathcal{B}$ from the left [resp. right] hand side can be described by the composition of the following elementary row [resp. column] operations:

(3) Multiply the i-th row [resp. column] by a non-zero element of K for $1 \le i \le n$.

(4) Replace the i-th row [resp. j-th column] by the sum of the i-th and j-th row [resp. column] , where $1 \le i < j \le n$.

If P_π denotes the permutation matrix associated with $\pi \in S_n$, we get the identity

(5) $\quad P_\pi I_{ij} P_\sigma^{-1} = I_{\pi(i),\sigma(j)} \qquad \text{for } \pi,\sigma \in S_n , \ 1 \le i,j \le n .$

Now we derive

(4.1) Theorem. *Bruhat Decomposition*

If K *is a field, then*

$$GL(n;K) = \bigcup_{\pi \in S_n} \mathcal{B} P_\pi \mathcal{B}$$

holds as a disjoint union.

Proof. Let $A = (a_{ij}) \in GL(n;K)$ and choose

$$k = \max\{ i \mid 1 \le i \le n , \ a_{i1} \neq 0 \} .$$

According to (3) and (4) we can find a matrix A_1 in $\mathcal{B}A\mathcal{B}$ with e_k as its first column and e_1' as its k-th row. In the second step apply the same procedure to the second column of A_1. After n steps we obtain a permutation matrix P_π.

Now let $\pi,\sigma \in S_n$ satisfy $\mathcal{B}P_\pi\mathcal{B} = \mathcal{B}P_\sigma\mathcal{B}$. Then $B = (b_{ij}) \in \mathcal{B}$ exists such that

$$P_\pi B P_\sigma^{-1} \in \mathcal{B}.$$

At first assume that there exists an i, $1 \le i \le n$, with $\pi(i) > \sigma(i)$. But then (5) yields that the $(\pi(i), \sigma(i))$-entry of $P_\pi B P_\sigma^{-1}$ equals $b_{ii} \neq 0$, which is a contradiction. On the other hand $\pi(i) \le \sigma(i)$ for $1 \le i \le n$ immediately leads to $\pi = \sigma$. □

The following assertion can be found in N. Bourbaki [9].

(4.2) Lemma. *Let* K *be a field,* $1 \le i < n$, *and* $\tau = \tau_{i,i+1}$. *Given* $\pi \in S_n$ *one has*

a) $$\mathcal{B}P_\tau\mathcal{B}P_\pi\mathcal{B} \subset \mathcal{B}P_\pi\mathcal{B} \cup \mathcal{B}P_{\tau\circ\pi}\mathcal{B},$$

b) $$\mathcal{B}P_\pi\mathcal{B}P_\tau\mathcal{B} \subset \mathcal{B}P_\pi\mathcal{B} \cup \mathcal{B}P_{\pi\circ\tau}\mathcal{B}.$$

Proof. a) It suffices to demonstrate

$$P_\tau\mathcal{B} \subset \mathcal{B}\mathcal{B}_\pi \cup \mathcal{B}P_\tau\mathcal{B}_\pi, \quad \text{where} \quad \mathcal{B}_\pi = P_\pi\mathcal{B}P_\pi^{-1}.$$

Now consider the subgroup

$$G_i = \left\{ \begin{pmatrix} I_1 & 0 & 0 \\ 0 & A & 0 \\ 0 & 0 & I_2 \end{pmatrix} \;\middle|\; I_1 = I^{(i-1)},\ A \in GL(2;K),\ I_2 = I^{(n-i-1)} \right\}$$

of $GL(n;K)$. A straightforward calculation yields

$$G_i\mathcal{B} = \mathcal{B}G_i.$$

by reduction to the obvious case $n = 2$. This implies $P_\tau\mathcal{B} \subset \mathcal{B}G_i$ and thus it suffices to demonstrate

$$G_i \subset (\mathcal{B}\cap G_i)(\mathcal{B}_\pi\cap G_i) \cup (\mathcal{B}\cap G_i)P_\tau(\mathcal{B}_\pi\cap G_i).$$

If we use the obvious isomorphism between G_i and $GL(2;K)$, clearly $\mathcal{B}\cap G_i$ is mapped onto $\mathcal{B}(2;K)$ and $\mathcal{B}_\pi\cap G_i$ onto

$$\mathcal{B}(2;K), \qquad \text{if } \pi(i) < \pi(i+1),$$

$$\begin{pmatrix} 0 & 1 \\ 1 & 0 \end{pmatrix}\mathcal{B}(2;K)\begin{pmatrix} 0 & 1 \\ 1 & 0 \end{pmatrix} = \left\{\begin{pmatrix} a & 0 \\ c & d \end{pmatrix} \in GL(2;K)\right\}, \qquad \text{if } \pi(i) > \pi(i+1).$$

In both cases the Bruhat decomposition for $n = 2$ in (4.1) yields the assertion.

b) Apply a) and use the map $GL(n;K) \longrightarrow GL(n;K)$, $A \longmapsto A^{-1}$. □

Given $\pi \in S_n$ we define

$$L(\pi) = \{ (i,j) \mid 1 \le i < j \le n \ , \ \pi(i) > \pi(j) \}$$

and call $\ell(\pi) := \# L(\pi)$ the *length* of π .

(4.3) Lemma. *Let* $\pi \in S_n$ *and* $\tau = \tau_{k,k+1}$, $0 < k < n$. *Then one has*

a) $$\det P_\pi = (-1)^{\ell(\pi)} = \operatorname{sign} \pi \ ,$$

b) $$\ell(\pi) = \ell(\pi^{-1}) \ ,$$

c) $$\ell(\pi \circ \tau) = \begin{cases} \ell(\pi) + 1 & \text{if } \pi(k) < \pi(k+1) \ , \\ \ell(\pi) - 1 & \text{if } \pi(k) > \pi(k+1) \ , \end{cases}$$

d) $$\ell(\tau \circ \pi) = \begin{cases} \ell(\pi) + 1 & \text{if } \pi^{-1}(k) < \pi^{-1}(k+1) \ , \\ \ell(\pi) - 1 & \text{if } \pi^{-1}(k) > \pi^{-1}(k+1) \ . \end{cases}$$

Proof. a) The assertion follows from the parts c) , d) , because the transpositions $\tau_{k,k+1}$, $0 < k < n$, generate S_n .
b) One easily checks

$$L(\pi^{-1}) = \{ (\pi(j), \pi(i)) \mid (i,j) \in L(\pi) \} \ .$$

c) One easily verifies

$$L(\pi \circ \tau) = \{ (\tau(i), \tau(j)) \mid (i,j) \in L(\pi) \} \cup \{ (k,k+1) \} , \quad \text{if } \pi(k) < \pi(k+1) \ ,$$

$$L(\pi \circ \tau) = \{ (\tau(i), \tau(j)) \mid (i,j) \in L(\pi) , (i,j) \neq (k,k+1) \} , \quad \text{if } \pi(k) > \pi(k+1) \ .$$

d) Apply b) and c) . □

Now we deal with the Hecke algebra attached to the pair $(\mathcal{B}, GL(n;K))$, where $K = GF(q)$ is a finite field with q elements.

(4.4) Lemma. *Let* $K = GF(q)$ *and* $\mathcal{B} = \mathcal{B}(n;K)$.
a) *Given* $\pi \in S_n$ *one has*

$$\operatorname{ind} \mathcal{B} P_\pi \mathcal{B} = q^{\ell(\pi)} \ .$$

b) *Let* $0 < k < n$. *Then a set of representatives of the right cosets relative to* $\mathcal{B}$ *in* $\mathcal{B} P_\tau \mathcal{B}$, $\tau = \tau_{k,k+1}$, *is given by*

$$P_\tau (I + a I_{k,k+1}) \ , \quad a \in K \ .$$

Proof. a) According to (5) one has

$$(*) \qquad \mathcal{B} \cap P_\pi^{-1} \mathcal{B} P_\pi = \{ A = (a_{ij}) \in \mathcal{B} \mid a_{ij} = 0 \ \text{ for } (i,j) \in L(\pi) \} ,$$

Using I(3.1) we get

$$\operatorname{ind} \mathcal{B}P_\pi\mathcal{B} = [\mathcal{B} : \mathcal{B} \cap P_\pi^{-1}\mathcal{B}P_\pi] = q^{\ell(\pi)} ,$$

since the order of $\mathcal{B}$ equals $(q-1)^n \, q^{n(n-1)/2}$.
b) A set of representatives of $(\mathcal{B} \cap P_\tau\mathcal{B}P_\tau)\backslash\mathcal{B}$ is given by $I + aI_{k,k+1}$, $a \in K$, according to (*) . Then

$$P_\tau\left(I + aI_{k,k+1}\right) , \quad a \in K ,$$

forms a set of representatives of $\mathcal{B}\backslash\mathcal{B}P_\tau\mathcal{B}$ due to I(3.1) . □

Now we compute the product of an arbitrary double coset with $\mathcal{B}P_\tau\mathcal{B}$, $\tau = \tau_{k,k+1}$, in the associated Hecke algebra.

(4.5) Proposition. *Let* $K = GF(q)$ *and* $\mathcal{B} = \mathcal{B}(n;K)$. *Given* $\pi \in S_n$ *and* $\tau = \tau_{k,k+1}$, *where* $0 < k < n$, *one has*

$$\mathcal{B}P_\pi\mathcal{B} \cdot \mathcal{B}P_\tau\mathcal{B} = \begin{cases} \mathcal{B}P_{\pi\circ\tau}\mathcal{B} , & \text{if } \pi(k) < \pi(k+1) , \\ q\,\mathcal{B}P_{\pi\circ\tau}\mathcal{B} + (q-1)\,\mathcal{B}P_\pi\mathcal{B} , & \text{if } \pi(k) > \pi(k+1) . \end{cases}$$

Proof. Assume $F = \mathbb{Q}$ without restriction. If $\pi(k) < \pi(k+1)$, then

$$\mathcal{B}P_\pi\mathcal{B} \cdot \mathcal{B}P_\tau\mathcal{B} = \mathcal{B}P_{\pi\circ\tau}\mathcal{B}$$

follows from (4.4), (4.3), and I(4.7) . If $\pi(k) > \pi(k+1)$, then

$$\mathcal{B}P_\pi\mathcal{B} \cdot \mathcal{B}P_\tau\mathcal{B} = t\,q\,\mathcal{B}P_{\pi\circ\tau}\mathcal{B} + (q-t)\,\mathcal{B}P_\pi\mathcal{B}$$

with some $t \in \{0,1, \dots ,q\}$ follows from (4.4) , (4.3) , (4.2) , and I(4.5) . Now apply I(4.4) and use the representatives in (4.4) . Then

$$P_{\pi\circ\tau}\left(P_\tau\left(I + aI_{k,k+1}\right)\right)^{-1} = P_\pi - aP_\pi I_{k+1,k} = \Big(\sum_{j=1}^{n} P_{\pi(j),j}\Big) - aI_{\pi(k+1),k}$$

belongs to $\mathcal{B}P_\pi\mathcal{B}$ for every $a \in K$ because of $\pi(k) > \pi(k+1)$ and the proof of (4.1) . Thus $t = 1$ follows. □

According to (4.3) we can reformulate (4.5) as

$$(6) \qquad \mathcal{B}P_\pi\mathcal{B} \cdot \mathcal{B}P_\tau\mathcal{B} = \begin{cases} \mathcal{B}P_{\pi\circ\tau}\mathcal{B} , & \text{if } \ell(\pi\circ\tau) > \ell(\pi) , \\ q\,\mathcal{B}P_{\pi\circ\tau}\mathcal{B} + (q-1)\,\mathcal{B}P_\pi\mathcal{B} , & \text{if } \ell(\pi\circ\tau) < \ell(\pi) . \end{cases}$$

The application of I(7.3) to (6) and (4.3) lead to

$$(7) \qquad \mathcal{B}P_\tau\mathcal{B} \cdot \mathcal{B}P_\pi\mathcal{B} = \begin{cases} \mathcal{B}P_{\tau\circ\pi}\mathcal{B} , & \text{if } \ell(\tau\circ\pi) > \ell(\pi) , \\ q\,\mathcal{B}P_{\tau\circ\pi}\mathcal{B} + (q-1)\,\mathcal{B}P_\pi\mathcal{B} , & \text{if } \ell(\tau\circ\pi) < \ell(\pi) . \end{cases}$$

We pick up some special cases. Therefore let

$$P_k := P_{\tau_{k,k+1}} , \quad 0 < k < n .$$

(4.6) Corollary. *Let* $K = GF(q)$ *and* $\mathcal{B} = \mathcal{B}(n;K)$.
a) *The Hecke algebra* $\mathcal{H}(\mathcal{B},GL(n;K))$ *is generated by the double cosets*

$$\mathcal{B}\ ,\ \mathcal{B}P_k\mathcal{B}\ ,\quad k = 1,\dots,n-1\ .$$

More precisely, given $\pi \in S_n$, *with* $\ell(\pi) = \ell > 0$, *there is a representation of the form*

(8) $$\mathcal{B}P_\pi\mathcal{B} = \mathcal{B}P_{i_1}\cdot\ldots\cdot\mathcal{B}P_{i_\ell}\mathcal{B} \quad \textit{with} \quad i_\nu \in \{1,2,\dots,n-1\}\ .$$

b) $$(\mathcal{B}P_k\mathcal{B})\cdot(\mathcal{B}P_\ell\mathcal{B}) = (\mathcal{B}P_\ell\mathcal{B})\cdot(\mathcal{B}P_k\mathcal{B}) = \mathcal{B}P_\pi\mathcal{B}$$

holds for $1 < k+1 < \ell < n$, *where* $\pi = \tau_{k,k+1} \circ \tau_{\ell,\ell+1}$.

c) $$(\mathcal{B}P_k\mathcal{B})\cdot(\mathcal{B}P_{k+1}\mathcal{B})\cdot(\mathcal{B}P_k\mathcal{B}) = (\mathcal{B}P_{k+1}\mathcal{B})\cdot(\mathcal{B}P_k\mathcal{B})\cdot(\mathcal{B}P_{k+1}\mathcal{B}) = \mathcal{B}P_\tau\mathcal{B}\ ,$$

holds for $0 < k < n-1$, *where* $\tau = \tau_{k,k+2}$.
d) *Given* $0 < k < n$ *one has*

$$(\mathcal{B}P_k\mathcal{B})^2 = q\mathcal{B} + (q-1)\mathcal{B}P_k\mathcal{B}\ .$$

Proof. a) S_n is generated by $\tau_{k,k+1}$, $k = 1,\dots,n$. More precisely each $\pi \in S_n$ with $\ell(\pi) = \ell$ possesses a representation

$$\pi = \tau_1 \circ\ldots\circ \tau_\ell\ ,\quad \text{where}\quad \tau_k \in \{\tau_{1,2},\dots,\tau_{n-1,n}\}\ .$$

due to (4.3) . Now apply (4.5) , (4.4) , and (4.1) in order to obtain (8) .
b), c), d) Use (4.5) for these particular double cosets. □

In this case we obtain another homomorphism besides ind .

(4.7) Lemma. *Let* $K = GF(q)$ *and* $\mathcal{B} = \mathcal{B}(n;K)$. *Then the map*

$$\mathrm{SIGN} : \mathcal{H}(\mathcal{B},GL(n;K)) \longrightarrow F\ ,$$
$$T = \sum_{\pi\in S_n} t(\pi)\mathcal{B}P_\pi\mathcal{B} \mapsto \sum_{\pi\in S_n} t(\pi)\,\mathrm{sign}\,\pi\ ,$$

is a homomorphism of the algebras.

Proof. The map SIGN is well-defined due to (4.1) and obviously linear. Then (4.5) yields

$$\mathrm{SIGN}\,(\mathcal{B}P_\pi\mathcal{B}\cdot\mathcal{B}P_k\mathcal{B}) = -\mathrm{sign}\ \pi = \mathrm{SIGN}\,(\mathcal{B}P_\pi\mathcal{B})\ \mathrm{SIGN}\,(\mathcal{B}P_k\mathcal{B}),\quad 0 < k < n\ .$$

Thus the multiplicativity follows from (4.6) . □

Note that there exists a unique element $\sigma \in S_n$ of maximal length

$$\sigma(j) = n+1-j\ ,\ 1 \le j \le n\ ,\ \ell(\sigma) = n(n-1)/2\ .$$

Furthermore there exist exactly $n-1$ elements of length $(n(n-1)/2)-1 = (n-2)(n+1)/2$, namely

$$\sigma\circ\tau_{k,k+1} = \tau_{n-k,n+1-k}\circ\sigma \quad , \quad k = 1,\dots,n-1 .$$

(4.8) Theorem. *Let* $K = GF(q)$ *and* $\mathcal{B} = \mathcal{B}(n;K)$. *Then the Hecke algebra* $\mathcal{H}(\mathcal{B},GL(n;K))$ *has the dimension* $n!$ *and is generated by*

$$\mathcal{B} \, , \, \mathcal{B}P_k\mathcal{B} \quad , \qquad k = 1,\dots,n-1 .$$

Now let $n \geq 3$. *Then* $\mathcal{H}(\mathcal{B},GL(n;K))$ *fails to be commutative.* $\mathcal{H}(\mathcal{B},GL(n;K))$ *is semi-simple if and only if*

$$\text{(9)} \qquad \operatorname{char} F \nmid q \prod_{j=2}^{n} \frac{q^j-1}{q-1} .$$

Proof. The first part is clear due to (4.6) and (4.1) . Moreover (4.5) yields the non-commutativity for $n \geq 3$ in view of

$$(\mathcal{B}P_2\mathcal{B})\cdot(\mathcal{B}P_1\mathcal{B}) \neq (\mathcal{B}P_1\mathcal{B})\cdot(\mathcal{B}P_2\mathcal{B}) .$$

The formula

$$\text{(10)} \qquad \# \, GL(n;K) = \prod_{j=0}^{n-1} (q^n - q^j)$$

can for instance be found in N. Jacobson [26] , p. 362 , or E. Artin [7] , p. 170 . For a proof of (10) note that we have q^n-1 possibilities for the first column of a matrix in $GL(n;K)$. If the first j columns of a matrix in $GL(n;K)$ are fixed, we have q^n-q^j possibilities for the next column, namely all the column vectors in K^n outside the j-dimensional subspace spanned by the first j columns.

Thus (10) leads to

$$[GL(n;K) : \mathcal{B}] = \prod_{j=2}^{n} \frac{q^j-1}{q-1} .$$

Hence (4.4) and II(1.6) imply that the Hecke algebra is semi-simple, whenever (9) holds. On the other hand it follows from II(1.3) that $\mathcal{H}(\mathcal{B},GL(n;K))$ fails to be semi-simple, if $\operatorname{char} F \mid \prod_{j=2}^{n} \frac{q^j-1}{q-1}$.

Therefore let us assume $\operatorname{char} F \mid q$. Then (6) and (7) imply that the set

$$\Big\{ \sum_{\pi\in S_n, \ell(\pi)\geq \ell} t(\pi)\,\mathcal{B}P_\pi\mathcal{B} \mid t(\pi)\in F \Big\}$$

is an ideal in $\mathcal{H}(\mathcal{B},GL(n;K))$ for $0 \leq \ell \leq n(n-1)/2$. Set $\bar{P}_k = P_{\sigma\circ\tau_{k,k+1}}$, $0 < k < n$. Then

$$\mathcal{J} = \Big\{ \sum_{k=1}^{n-1} t_k\big(\mathcal{B}\bar{P}_k\mathcal{B} + \mathcal{B}P_\sigma\mathcal{B}\big) \mid t_k \in F \Big\}$$

$$= \Big\{ T = \sum_{\pi\in S_n, \ell(\pi)\geq (n-2)(n+1)/2} t(\pi)\,\mathcal{B}P_\pi\mathcal{B} \in \mathcal{H}(\mathcal{B},GL(n;K)) \mid \operatorname{SIGN} T = 0 \Big\}$$

proves to be an ideal in $\mathcal{H}(\mathcal{B},GL(n;K))$ due to (4.7). Using (4.7), (4.5) and char F | q we get

$$\mathcal{B}P_\sigma\mathcal{B}\cdot T = T\cdot\mathcal{B}P_\sigma\mathcal{B} = (\mathrm{SIGN}\ T)\mathcal{B}P_\sigma\mathcal{B} \qquad \text{for } T\in\mathcal{H}(\mathcal{B},GL(n;K)),$$

$$\text{(11)}\qquad \mathcal{B}\bar{P}_k\mathcal{B}\cdot\mathcal{B}P_\ell\mathcal{B} = \begin{cases}\mathcal{B}P_\sigma\mathcal{B} & \text{if } k=\ell,\\ -\mathcal{B}\bar{P}_k\mathcal{B} & \text{if } k\neq\ell.\end{cases}$$

Now write $\mathcal{B}\bar{P}_k\mathcal{B}$ in the form (8). If a double coset $\mathcal{B}P_\pi\mathcal{B}$ is represented as a prod- of the $\mathcal{B}P_i\mathcal{B}$ in the form (8) with $i\neq\ell$, we conclude

$$\ell(\pi) \le \big(\ell(\ell-1)+(n-\ell)(n-1-\ell)\big)/2 < \big(n(n-1)/2\big)-1$$

in view of $n\ge 3$. Hence we know that each $\mathcal{B}P_\ell\mathcal{B}$ occurs in any representation (8) of the element $\mathcal{B}\bar{P}_k\mathcal{B}$. Then (11) yields

$$\mathcal{B}\bar{P}_k\mathcal{B}\cdot\mathcal{B}\bar{P}_\ell\mathcal{B} = \pm\mathcal{B}P_\sigma\mathcal{B}.$$

In view of SIGN $T=0$ for $T\in\mathcal{J}$, we conclude $T_1\cdot T_2=0$ for $T_1,T_2\in\mathcal{J}$. Thus $\mathcal{J}$ is a nilpotent ideal in $\mathcal{H}(\mathcal{B},GL(n;K))$. □

S_n is the so-called *Weyl group* of GL(n;K). There exists a connection between the Hecke algebra $\mathcal{H}(\mathcal{B},GL(n;K))$ and the group algebra of S_n. If we formally set $q=1$ in the products in (4.5) and (4.6), then $\mathcal{B}P_\pi\mathcal{B}\mapsto\pi$ exactly yields the group algebra of S_n. A generalization of this phenomenon for a wider class of groups is described in

(4.9) Remark. Let the assumptions of (4.6) hold. N. Iwahori [24] showed that the Hecke algebra $\mathcal{H}(\mathcal{B},GL(n;K))$ can be characterized by generators $\mathcal{B}P_k\mathcal{B}$, $k=1,\dots,n-1$, and the defining relations

(i) $(\mathcal{B}P_k\mathcal{B})\cdot(\mathcal{B}P_\ell\mathcal{B}) = (\mathcal{B}P_\ell\mathcal{B})\cdot(\mathcal{B}P_k\mathcal{B})$ for $1<k+1<\ell<n$,

(ii) $(\mathcal{B}P_k\mathcal{B})\cdot(\mathcal{B}P_{k+1}\mathcal{B})\cdot(\mathcal{B}P_k\mathcal{B}) = (\mathcal{B}P_{k+1}\mathcal{B})\cdot(\mathcal{B}P_k\mathcal{B})\cdot(\mathcal{B}P_{k+1}\mathcal{B})$ for $0<k<n-1$,

(iii) $(\mathcal{B}P_k\mathcal{B})^2 = q\mathcal{B}+(q-1)\mathcal{B}P_k\mathcal{B}$ for $0<k<n$.

This can be generalized to an arbitrary finite group G with a (B,N)-pair in place of GL(n;K). Then the Hecke algebra $\mathcal{H}(B,G)$ has an analogous presentation with generators and relations. In this situation a *generic algebra* (cf. [14], [17]) is introduced. Now the Hecke algebra $\mathcal{H}(B,G)$ and the group algebra of the Weyl group W of G are obtained as specialized algebras of the generic algebra. As specialized algebras of the same generic algebra the numerical invariants of the Hecke algebra and the group algebra are equal, i.e., $\mathcal{H}(B,G)$ over $\mathbb{C}$ and $\mathbb{C}[W]$ are isomorphic.

More generally one can consider a parabolic subgroup P_J of G in place of B. Denote the induced parabolic subgroup of W by W_J. Then C.W. Curtis, N. Iwahori, and R. Kilmoyer [15] showed that the Hecke algebras $\mathcal{H}(P_J,G)$ and $\mathcal{H}(W_J,W)$ over $\mathbb{C}$ are isomorphic.

(4.10) Exercises. a) Show that the Hecke algebra $\mathcal{H}(\mathcal{B},GL(2;K))$ is two-dimensional, commutative and moreover semi-simple, if and only if char $F \nmid (q+1)$.
b) Determine the center of $\mathcal{H}(\mathcal{B},GL(3;K))$.
c) Show that $\mathcal{H}(\mathcal{B},GL(3;K))$ is isomorphic to

$$F \times F \times Mat(2;F) ,$$

whenever F is real closed or algebraically closed with char $F = 0$. (If F is arbitrary, confer Curtis/Reiner [17] , (67.14) .)
d) Show that each $\mathcal{B}P_\pi\mathcal{B}$ is a unit in $\mathcal{H}(\mathcal{B},GL(n;K))$, if and only if char $F \nmid q$.
e) Consider $0 < m < n$ and $S_m \subset S_n$ as in section 2 . Show that the set

$$\Big\{ T = \sum_{\pi\in S_m} t(\pi)\mathcal{B}P_\pi\mathcal{B} \mid t(\pi) \in F \Big\} \ , \ \mathcal{B} = \mathcal{B}(n;K) \ ,$$

is a subalgebra of $\mathcal{H}(\mathcal{B},GL(n;K))$, which is isomorphic to $\mathcal{H}(\mathcal{B},GL(m;K))$.
f) Let $\pi \in S_n$, $\pi \neq id$, and assume that char $F = 0$. Show that the following assertions are equivalent:

(i) $\pi = \tau_{j,j+1}$ for some j , $0 < j < n$.

(ii) $\mathcal{B} \cup \mathcal{B}P_\pi\mathcal{B}$ is a subgroup of $GL(n;K)$.

(iii) $F\mathcal{B} + F\mathcal{B}P_\pi\mathcal{B}$ is a subalgebra of $\mathcal{H}(\mathcal{B},GL(n;K))$.

5. A Maximal Parabolic Subgroup

In this section we pick up the case of a maximal parabolic subgroup. Given $m,n \in \mathbb{N}$ consider

$$\begin{aligned}\mathcal{P} = \mathcal{P}(m,n;K) &= \Big\{M = \begin{pmatrix} * & * \\ 0 & * \end{pmatrix} \in GL(m+n;K) \mid 0 = 0^{(n,m)}\Big\} \\ &= \Big\{M = \begin{pmatrix} A & B \\ 0 & D \end{pmatrix} \mid A \in GL(m;K) \ , \ D \in GL(n;K) \ , \ B \in Mat(m,n;K)\Big\},\end{aligned}$$

whenever K is an arbitrary field. Since the subgroups $\mathcal{P}(m,n;K)$ and $\mathcal{P}(n,m;K)'$ are conjugate in $GL(m+n;K)$, we will always assume

$$m \le n \ .$$

There are two natural Hecke algebras involving $\mathcal{P}$, where $K = GF(q)$. First of all consider $G = \mathcal{P}$ and the subgroup

$$S = GL(m;K) \times GL(n;K) = \Big\{\begin{pmatrix} A & 0 \\ 0 & D \end{pmatrix} \mid A \in GL(m;K) , D \in GL(n;K)\Big\} .$$

The attached Hecke algebra is commutative, because $\mathcal{P}$ is the semi-direct product of S and $Mat(m,n;K)$ such that we can apply I(7.4) . A set of representatives of the right cosets is given by the matrices

(1) $$\begin{pmatrix} I & B \\ 0 & I \end{pmatrix} , \ B \in Mat(m,n;K) .$$

Due to the usual normal form and $m \le n$ there are exactly $m+1$ double cosets given by rank conditions namely

(2) $$\mathcal{R}_k = \bigcup_{\substack{B \in Mat(m,n;K) \\ \text{rank } B = k}} \mathcal{P}\begin{pmatrix} I & B \\ 0 & I \end{pmatrix}, \quad k = 0, \ldots, m .$$

Here clearly $\mathcal{R}_0 = S$ holds. Obviously $\mathcal{R}_k$ contains a unique representative of the form

$$\begin{pmatrix} I & B \\ 0 & I \end{pmatrix}, \quad B = \begin{pmatrix} I^{(k)} & 0 \\ 0 & 0 \end{pmatrix} .$$

(5.1) Proposition. *Let* $K = GF(q)$, $1 \le m \le n$. *Then the Hecke algebra under consideration* $\mathcal{H}(GL(m;K) \times GL(n;K), \mathcal{P}(m,n;K))$ *is commutative and has the dimension* $m+1$. *One has the following products*

$$\begin{aligned}(q-1)\, \mathcal{R}_k \cdot \mathcal{R}_1 &= q^{2k-2} (q^{n+1-k} -1)(q^{m+1-k} -1)\, \mathcal{R}_{k-1} \\ &\quad + (q^k -1)(q^m + q^n - q^k - q^{k-1} -1)\, \mathcal{R}_k + q^k (q^{k+1} -1)\, \mathcal{R}_{k+1}\end{aligned}$$

for $0 < k < m$ *as well as*

$$(q-1)\mathcal{R}_m \cdot \mathcal{R}_1 = q^{2m-2} (q^{n+1-m} -1)(q-1)\, \mathcal{R}_{m-1} + (q^m -1)(q^n - q^{m-1} -1)\mathcal{R}_m .$$

Proof. We observe that ab' runs $(q-1)$-times through a set of all matrices $B \in Mat(m,n;K)$ with rank $B = 1$, whenever a runs through $K^m \setminus \{0\}$ and b runs through $K^n \setminus \{0\}$. Given $0 \le \ell \le n$ elementary Linear Algebra shows that among the matrices

$$C = \begin{pmatrix} I^{(\ell)} & 0 \\ 0 & 0 \end{pmatrix} - ab',$$

we have $q^{\ell-1}(q^\ell -1)$ ones with rank $C = \ell-1$. There are $q^{2\ell}(q^{n-\ell}-1)(q^{m-\ell}-1)$ matrices C with rank $C = \ell+1$. Hence $(q^\ell-1)(q^m+q^n-q^\ell-q^{\ell-1}-1)$ matrices C have the rank ℓ. Using (1) and (2) the assertion follows from I(4.4i). □

As an obvious consequence we conclude that the Hecke algebra coincides with $F[\mathcal{R}_1]$, whenever

$$\text{char}\, F \nmid q \prod_{k=2}^{m} \frac{q^k -1}{q-1} ,$$

where the vacuous product is of course defined to be 1. Moreover (2) yields

$$\text{ind}\, \mathcal{R}_k = \#\{ B \in Mat(m,n;K) \mid \text{rank } B = k \} .$$

Using the maximal rank decomposition for B we obtain

$$\text{ind}\, \mathcal{R}_k = \prod_{j=0}^{k-1} \frac{(q^m - q^j)(q^n - q^j)}{q^k - q^j}, \quad k = 1, \ldots, m .$$

But there is also another natural construction. In this case we deal with the Hecke algebra $\mathcal{H}(\mathcal{P}(m,n;K), GL(m+n;K))$. Given $0 \le k \le m$ consider the matrix

$$P_k = P_\pi \quad , \quad \pi = \tau_{1,m+1} \circ \ldots \circ \tau_{k,m+k} \, ,$$

$$P_k = \begin{pmatrix} 0 & 0 & I_1 & 0 \\ 0 & I_2 & 0 & 0 \\ I_1 & 0 & 0 & 0 \\ 0 & 0 & 0 & I_3 \end{pmatrix} \, , \quad I_1 = I^{(k)} \; , \; I_2 = I^{(m-k)} \; , \; I_3 = I^{(n-k)} \, .$$

The Bruhat decomposition for the parabolic subgroup $\mathcal{P}$ is derived in

(5.2) Lemma. *Let* K *be a field and* $1 \le m \le n$. *Then* $GL(m+n;K)$ *decomposes into* $m+1$ *double cosets relative to* $\mathcal{P} = \mathcal{P}(m,n;K)$, *namely*

$$\mathcal{P}P_k\mathcal{P} \, , \; k = 0, \ldots ,m \, .$$

Proof. In each double coset we obtain a representative P_k due to (4.1) combined with the results of section 3 . Let

$$M = \begin{pmatrix} A & B \\ C & D \end{pmatrix} \in GL(m+n;K) \; , \; A \in Mat(m;K) \; , \; D \in Mat(n;K) \, .$$

Since the ranks of the C-blocks coincide for the matrices in a double coset relative to $\mathcal{P}$, the double cosets $\mathcal{P}P_0\mathcal{P}, \ldots , \mathcal{P}P_m\mathcal{P}$ are mutually distinct. □

In the next step we determine a set of representatives of the right cosets in each double coset.

(5.3) Proposition. *Let* K *and* $1 \le k \le m \le n$. *A set of representatives of the right cosets relative to* $\mathcal{P} = \mathcal{P}(m,n;K)$ *in* $\mathcal{P}P_k\mathcal{P}$ *is given by*

$$P_k\begin{pmatrix} I & B \\ 0 & I \end{pmatrix}\begin{pmatrix} A'^{-1} & 0 \\ 0 & D \end{pmatrix} \, ,$$

where $B = \begin{pmatrix} B_1 & 0 \\ 0 & 0 \end{pmatrix} \in Mat(m,n;K)$, $B_1 \in Mat(k;K)$, A *runs through a set of representatives of* $\mathcal{P}(k,m-k;K)\backslash GL(m;K)$ *and* D *of* $\mathcal{P}(k,n-k;K)\backslash GL(n;K)$.

Proof. Given $M \in \mathcal{P}$ we obtain

$$\bar{M} P_k = P_k M$$

for some $\bar{M} \in \mathcal{P}$ if and only if $\bar{M} = P_k M P_k^{-1}$ and M has the form

$$M = \begin{pmatrix} A_1 & 0 & 0 & B_2 \\ A_3 & A_4 & B_3 & B_4 \\ 0 & 0 & D_1 & D_2 \\ 0 & 0 & 0 & D_4 \end{pmatrix} \, , \qquad \begin{array}{l} A_1 \, , \, D_1 \in GL(k;K) \; , \\ A_4 \in GL(m-k;K) \; , \\ D_4 \in GL(n-k;K) \; . \end{array}$$

This is equivalent to

$$M = \begin{pmatrix} I & B \\ 0 & I \end{pmatrix}\begin{pmatrix} A & 0 \\ 0 & D \end{pmatrix}, \; B = \begin{pmatrix} 0^{(k)} & * \\ * & * \end{pmatrix} \in Mat(m,n;K) \, , \; A \in \mathcal{P}(k,m-k;K)' \, , \; D \in \mathcal{P}(k,n-k;K) \, .$$

Since each $M \in \mathcal{P}$ can be written in the form

$$M = \begin{pmatrix} I & B \\ 0 & I \end{pmatrix}\begin{pmatrix} A & 0 \\ 0 & D \end{pmatrix}, \quad B \in \mathrm{Mat}(m,n;K)\ ,\ A \in GL(m;K)\ ,\ D \in GL(n;K)\ ,$$

the assertion follows. □

In the next step we determine a set of representatives of the right cosets relative to $\mathcal{P}(1,n-1;K)$ in $GL(n;K)$.

(5.4) Lemma. *A set of representatives of* $\mathcal{P}(1,n-1;K)\backslash GL(n;K)$ *is given by the matrices* A_i^{-1}, $i = 1,\dots,n$, *where*

$$A_i = \begin{pmatrix} 0 & 0 & I_2 \\ 1 & 0 & 0 \\ a & I_1 & 0 \end{pmatrix}, \quad a \in K^{n-i}\ ,\quad I_1 = I^{(n-i)},\ I_2 = I^{(i-1)}\ .$$

Proof. We can equivalently prove that the A_i form a set of representatives of the left cosets $GL(n;K)/\mathcal{P}(1,n-1;K)$. But this is clear because of

$$\mathcal{P}(1,n-1;K) = \left\{ \begin{pmatrix} \alpha & b' \\ 0 & V \end{pmatrix} \mid 0 \neq \alpha \in K\ ,\ V \in GL(n-1;K)\ ,\ b \in K^{n-1} \right\}.$$

Finally $A,B \in GL(n;K)$ have the same first column if and only if

$$A^{-1}B = \begin{pmatrix} 1 & * \\ 0 & * \end{pmatrix}.$$

□

For a Hecke algebra we again need a finite field $K = GF(q)$. In view of the formula (10) in section 4 for $\# GL(n;K)$ we obtain

$$[GL(m+n;K) : \mathcal{P}(m,n;K)] = \prod_{j=1}^{m} \frac{q^{n+j}-1}{q^j-1}\ , \tag{3}$$

$$\mathrm{ind}\ \mathcal{P}P_k\mathcal{P} = q^{k^2} \prod_{j=0}^{k-1} \frac{(q^{m-j}-1)(q^{n-j}-1)}{(q^{k-j}-1)^2}\ ,\quad k = 0,\dots,m\ . \tag{4}$$

The Hecke algebra $\mathcal{H}(\mathcal{P}(m,n;K),GL(m+n;K))$ has the dimension $m+1$ for $1 \le m \le n$ and turns out to be commutative by virtue of I(7.3) and (5.2).

(5.5) Proposition. *Let* $K = GF(q)$ *and* $1 \le m \le n$. *Given* $0 < k < m$ *one has the following products*

$$\begin{aligned}(q-1)^2\, \mathcal{P}P_k\mathcal{P}\cdot\mathcal{P}P_1\mathcal{P} &= q^{2k-1}(q^{n+1-k}-1)(q^{m+1-k}-1)\,\mathcal{P}P_{k-1}\mathcal{P} \\ &+ (q^k-1)(q^{n+1}+q^{m+1}-q^{k+1}-q^k-q+1)\,\mathcal{P}P_k\mathcal{P} + (q^{k+1}-1)^2\,\mathcal{P}P_{k+1}\mathcal{P}\end{aligned}$$

as well as

$$\begin{aligned}(q-1)^2\, \mathcal{P}P_m\mathcal{P}\cdot\mathcal{P}P_1\mathcal{P} &= q^{2m-1}(q^{n+1-m}-1)(q-1)\,\mathcal{P}P_{m-1}\mathcal{P} \\ &+ (q^m-1)(q^{n+1}-q^m-q+1)\,\mathcal{P}P_m\mathcal{P}\ .\end{aligned}$$

Proof. Given $0 \le \ell \le m$ compute $P_\ell M^{-1}$, where M runs through the set of representatives of $\mathcal{P}\backslash\mathcal{P}P_1\mathcal{P}$ described in (5.3). Using (5.4) the C-block of $P_\ell M^{-1}$ is given by

$$C = \begin{pmatrix} \beta a_1 & a_2 & \dots & a_m \\ d & 0 & \dots & 0 \end{pmatrix},$$

where $\beta \in K$, $a_1, \dots, a_m$ are the columns of A_i^* consisting of the first ℓ rows of the matrix

$$A_i'^{-1} = \begin{pmatrix} 0 & 0 & I \\ 1 & -x' & 0 \\ 0 & I & 0 \end{pmatrix}, \quad x = (x_{i+1}, \dots, x_m)' \in K^{m-i}, \ i = 1, \dots, m,$$

and d denotes the column vector consisting of the last $n-\ell$ entries of the first column of

$$D_j = \begin{pmatrix} 0 & 0 & I \\ 1 & 0 & 0 \\ y & I & 0 \end{pmatrix}, \quad y = (y_{j+1}, \dots, y_n)' \in K^{n-j}, \ j = 1, \dots, n.$$

Since A_i^* has rank ℓ, we obtain

$$\ell - 1 \le \operatorname{rank} C \le \ell + 1 .$$

But $\operatorname{rank} C = \ell - 1$ is equivalent to $\beta = 0$, $d = 0$, $1 \le i \le \ell$, and $x_{\ell+1} = \dots = x_m = 0$. Hence this case occurs

$$\left(\frac{q^\ell - 1}{q-1}\right)^2$$

times. Next $\operatorname{rank} C = \ell + 1$ is equivalent to $d \neq 0$ as well as

$$1 \le i \le \ell, \quad (x_{\ell+1}, \dots, x_m) \neq (0, \dots, 0) \ \text{or} \ \ell < i \le m .$$

Thus this case occurs

$$q^{2\ell+1} \frac{q^{n-\ell}-1}{q-1} \frac{q^{m-\ell}-1}{q-1}$$

times. Because of (4) there remain

$$\frac{q^\ell - 1}{(q-1)^2} \left(q^{n+1} + q^{m+1} - q^{\ell+1} - q^\ell - q + 1\right)$$

cases with $\operatorname{rank} C = \ell$.

Now the assertion follows immediately from the description (i) of the product in I(4.4). □

Here we can remark the analogue of section 4. Set formally $q = 1$ in the products of (5.5), then $\mathcal{P}P_k\mathcal{P} \mapsto C_k$ exactly yields the Hecke algebra $\mathcal{H}(S_m \times S_n, S_{m+n})$ (cf. (3.2)).

Summarizing we have

(5.6) Theorem. *Let* $K = GF(q)$ *and assume that* $1 \le m \le n$. *Then the Hecke algebra* $\mathcal{H} = \mathcal{H}(\mathcal{P}(m,n;K),GL(m+n;K))$ *is commutative with dimension* $m+1$. *If* $\operatorname{char} F \nmid \prod_{k=2}^{m} \frac{q^k-1}{q-1}$, *one has*

$$\mathcal{H} = F[\mathcal{P}P_1\mathcal{P}] .$$

We know that the Hecke algebra fails to be semi-simple, whenever char F divides the index (3) . But we can also deal with the case $\operatorname{char} F \mid q$.

(5.7) Lemma. *Let* $K = GF(q)$, $\operatorname{char} F \mid q$, $\mathcal{H} = \mathcal{H}(\mathcal{P}(m,n;K),GL(m+n;K))$ *and* $2 \le m \le n$. *Then* rad $\mathcal{H}$ *is the principal ideal generated by* $\mathcal{P}P_2\mathcal{P}$ *and* $\mathcal{H}/\mathrm{rad}\,\mathcal{H}$ *is isomorphic to* $F \times F$.

Proof. (5.6) yields $\mathcal{H} = F[\mathcal{P}P_1\mathcal{P}]$. Moreover (5.5) implies

$$\mathcal{P}P_k\mathcal{P} \cdot \mathcal{P}P_1\mathcal{P} = -\mathcal{P}P_k\mathcal{P} + \mathcal{P}P_{k+1}\mathcal{P} ,\ 1 \le k < m ,\quad \mathcal{P}P_m\mathcal{P} \cdot \mathcal{P}P_1\mathcal{P} = -\mathcal{P}P_m\mathcal{P} .$$

Thus one easily checks

$$\mathcal{P}P_2\mathcal{P} = (\mathcal{P}P_1\mathcal{P})^2 + \mathcal{P}P_1\mathcal{P} ,$$

$$\mathcal{P}P_k\mathcal{P} \cdot \mathcal{P}P_2\mathcal{P} = \mathcal{P}P_{k+2}\mathcal{P} - \mathcal{P}P_{k+1}\mathcal{P} \qquad \text{for } 1 \le k \le m-2 ,$$

$$\mathcal{P}P_{m-1}\mathcal{P} \cdot \mathcal{P}P_2\mathcal{P} = -\mathcal{P}P_m\mathcal{P} ,\qquad \mathcal{P}P_m\mathcal{P} \cdot \mathcal{P}P_2\mathcal{P} = 0 ,\quad (\mathcal{P}P_2\mathcal{P})^m = 0 .$$

Hence the ideal $\mathcal{J}$ generated by $\mathcal{P}P_2\mathcal{P}$ is nilpotent and equal to

$$\mathcal{J} = F\,\mathcal{P}P_2\mathcal{P} + F\,\mathcal{P}P_3\mathcal{P} + \ldots + F\,\mathcal{P}P_m\mathcal{P} .$$

Since $(\mathcal{P} + \mathcal{P}P_1\mathcal{P}) + \mathcal{J}$, $-\mathcal{P}P_1\mathcal{P} + \mathcal{J}$ is a complete system of orthogonal idempotents in $\mathcal{H}/\mathcal{J}$, we conclude

$$\mathcal{H}/\mathcal{J} \cong F \times F .$$

Thus also $\mathcal{J} = \mathrm{rad}\,\mathcal{H}$ follows. □

Moreover we obtain a precise description in the case $m = 2$.

(5.8) Lemma. *Let* $K = GF(q)$, $n \ge 2$ *and* $\mathcal{H} = \mathcal{H}(\mathcal{P}(2,n;K),GL(n+2;K))$. *Then* $\mathcal{H}$ *over* F *is semi-simple if and only if*

$$\operatorname{char} F \nmid q \,\frac{(q^n-1)}{(q-1)} \frac{(q^{n+1}-1)}{(q-1)} \frac{(q^{n+2}-1)}{(q^2-1)} .$$

In this case $\mathcal{H}$ *is isomorphic to* $F \times F \times F$.

Proof. According to II(1.3) , (3) , and (5.7) $\mathcal{H}$ fails to be semi-simple, whenever

$$\operatorname{char} F \mid r = \frac{(q^{n+1}-1)(q^{n+2}-1)}{(q-1)\ \ (q^2-1)} \quad \text{or} \quad \operatorname{char} F \mid q .$$

If $\operatorname{char} F \mid \frac{q^n-1}{q-1}$, then $q\,\mathcal{P}P_1\mathcal{P} + (q+1)\mathcal{P}P_2\mathcal{P}$ turns out to be nilpotent due to (5.5).

Now let $\operatorname{char} F \mid (q+1)$. If n is even, then $\mathcal{P}P_1\mathcal{P}$ is nilpotent. If n is odd as well as $\operatorname{char} F \nmid r$, then $(\mathcal{P}P_1\mathcal{P})^2 = -(\mathcal{P}P_1\mathcal{P})$ follows and $\frac{1}{r}T_0$, $-\mathcal{P}P_1\mathcal{P}$, $\mathcal{P} + \mathcal{P}P_1\mathcal{P} - \frac{1}{r}T_0$ proves to be a complete system of orthogonal idempotents.

Now assume that $\operatorname{char} F \nmid (q+1)$, hence $\mathcal{H} = F[\mathcal{P}P_1\mathcal{P}]$ due to (5.6). Considering the left multiplication by $\mathcal{P}P_1\mathcal{P}$, we observe that $\mathcal{H}$ is isomorphic to $F[A]$, where

$$A = \begin{pmatrix} 0 & a & 0 \\ 1 & b & a-b-1 \\ 0 & c & a-c \end{pmatrix}, \quad a = q(q+1)\,\frac{q^n-1}{q-1}, \quad b = q\,\frac{q^n-1}{q-1} + q^2 - 1, \quad c = (q+1)^2,$$

according to (5.5). But A has the eigenvalues

$$q(q+1)\,\frac{q^n-1}{q-1}, \quad -(q+1), \quad q\,\frac{q^n-1}{q-1} - (q+1).$$

They are mutually different, whenever

$$\operatorname{char} F \nmid \quad q(q+1)\,\frac{q^n-1}{q-1}\,\frac{q^{n+1}-1}{q-1}\,\frac{q^{n+2}-1}{q-1}.$$

Thus $\mathcal{H}$ is isomorphic to $F \times F \times F$ in this case. □

(5.9) Remark. According to the case of an arbitrary parabolic subgroup of the general linear group, which was dealt with by C.W. Curtis, N. Iwahori, and R. Kilmoyer [15], we obtain an isomorphism between $\mathcal{H}(\mathcal{P},GL(m+n;K))$, $\mathcal{P} = \mathcal{P}(m,n;K)$, and $\mathcal{H}(S_m \times S_n, S_{m+n})$ over $\mathbb{C}$. In our particular case this is trivial, because both Hecke algebras are isomorphic to $m+1$ copies of F, whenever F is real closed or algebraically closed with $\operatorname{char} F = 0$ (cf. II(1.6), II(5.3), (3.3), (5.6)).

Here we give a simple application of the results of C.W. Curtis, N. Iwahori, and R. Kilmoyer in [15]. A *parabolic subgroup* of $GL(n;K)$ is a proper subgroup of $GL(n;K)$, which contains a conjugate of $\mathcal{B}(n;K)$. The minimal parabolic subgroups are therefore the conjugates of $\mathcal{B}(n;K)$. The maximal parabolic subgroups are exactly the conjugates of $\mathcal{P}(j,n-j;K)$, $j = 1,\dots,n-1$.

(5.10) Theorem. *Let* $\mathcal{P}$ *be a parabolic subgroup of* $GL(n;K)$, $n > 1$, $K = GF(q)$, *and* F *a field with* $\operatorname{char} F = 0$. *The Hecke algebra* $\mathcal{H}(\mathcal{P},GL(n;K))$ *over* F *is commutative if and only if* $\mathcal{P}$ *is a maximal parabolic subgroup.*

Proof. If $\mathcal{P}$ is a maximal parabolic subgroup, apply (5.6). If $\mathcal{P}$ fails to be maximal, then (3.5) and Theorem 3.1 in [15] imply that the associated Hecke algebra is non-commutative. □

Note that the theorem includes the case $n = 2$, where $\mathcal{B}$ is the only parabolic subgroup up to conjugates.

(5.11) Exercises. a) Show that for $n \geq 2$ the Hecke algebra $\mathcal{H}(GL(2;K) \times GL(n;K), \mathcal{P}(2,n;K))$ is semi-simple if and only if char $F \nmid q$. In this case the Hecke algebra is isomorphic to $F \times F \times F$. Determine the radical in the non-semi-simple case.

b) Determine the radical of the Hecke algebra in the non-semi-simple cases of (5.8).

c) Let $K = GF(q)$. Show that the Hecke algebras $\mathcal{H}(GL(m;K) \times GL(n;K), \mathcal{P}(m,n;K))$ and $\mathcal{H}(GL(n;K) \times GL(m;K), \mathcal{P}(n,m;K))$ are isomorphic for $m,n \in \mathbb{N}$ (cf. (5.1)). Moreover show that the Hecke algebras $\mathcal{H}(\mathcal{P}(m,n;K), GL(m+n;K))$ as well as $\mathcal{H}(\mathcal{P}(n,m;K), GL(m+n;K))$ are isomorphic for arbitrary $m,n \in \mathbb{N}$.

d) Fix positive integers ℓ,m,n and consider the parabolic subgroup

$$\mathcal{P} = \left\{ \begin{pmatrix} A & B & C \\ 0 & D & E \\ 0 & 0 & F \end{pmatrix} \in GL(\ell+m+n;K) \mid A = A^{(\ell)}, \ D = D^{(m)}, \ F = F^{(n)} \right\}$$

of $GL(\ell+m+n;K)$. Determine the Bruhat decomposition with respect to $\mathcal{P}$ as well as $\dim \mathcal{H}(\mathcal{P}, GL(\ell+m+n;K))$, whenever $K = GF(q)$.

Remarks

The example in section 1 was described by T. Kleine Schonefeld [29]. The other examples should illustrate the more general theory, which of course does not have our explicit character. For the general theory of Hecke algebras for finite groups with a (B,N)-pair the reader is referred to C.W. Curtis [14], C.W. Curtis, N. Iwahori, and R. Kilmoyer [15], C.W. Curtis and I. Reiner [16], [17], N. Iwahori [24], or H. Matsumoto [46].

Chapter IV

The Hecke Algebra Associated with $SL(2;\mathbb{Z})$

In this chapter we describe the Hecke algebra associated with $SL(2;\mathbb{Z})$. Although the results are also contained in the next chapter, where we deal with the Hecke algebra associated with $GL(n;\mathbb{Z})$, a separate treatment of the case $n=2$ seems to be justified. In the simplest applications to elliptic modular forms and Maass wave forms the Hecke algebra associated with $SL(2;\mathbb{Z})$ is used. Moreover all the proofs can be given more explicitly, which increases the transparency. We will also deal with congruence subgroups. A particular subalgebra of the Hecke algebra associated with $\Gamma_0(q)$ will be described.

Let us fix some notations. The group

$$\Gamma := SL(2;\mathbb{Z}) := \{U \in Mat(2;\mathbb{Z}) \mid \det U = 1\}$$

is called the *modular group*. Given $\ell \in \mathbb{N}$ consider the determinantal set

$$M(\ell) := \{A \in Mat(2;\mathbb{Z}) \mid \det A = \ell\},$$

where clearly $M(1) = \Gamma$. Then

$$M := \bigcup_{\ell=1}^{\infty} M(\ell) = \{A \in Mat(2;\mathbb{Z}) \mid \det A > 0\}$$

becomes a monoid, which is contained in the group

$$GL^+(2;\mathbb{Q}) := \{A \in Mat(2;\mathbb{Q}) \mid \det A > 0\}.$$

We will show that $(\Gamma, GL^+(2;\mathbb{Q}))$ is a Hecke pair. The aim of this chapter is to describe the structure of the Hecke algebra $\mathcal{H} = \mathcal{H}(\Gamma, M)$ over $R = \mathbb{Z}$.

1. Right Coset Decompositions

First of all we need a good description of representatives of the right cosets relative to Γ in M.

(1.1) Lemma. *Given* $0 \neq \binom{a}{c} \in \mathbb{Z}^2$ *there exists* $U \in \Gamma$ *satisfying*

$$U\binom{a}{c} = \binom{\delta}{0}, \quad \delta = \text{g.c.d.}\,(a,c).$$

Proof. We may replace $\binom{a}{c}$ by $\frac{1}{\delta}\binom{a}{c} \in \mathbb{Z}^2$ and therefore assume g.c.d. (a,c) = 1 without restriction. Hence there exist $b, d \in \mathbb{Z}$ such that $ad - bc = 1$. Now choose

$$U = \begin{pmatrix} a & b \\ c & d \end{pmatrix}^{-1} = \begin{pmatrix} d & -b \\ -c & a \end{pmatrix} \in \Gamma. \qquad \square$$

This lemma is applied in order to prove

(1.2) Proposition. *Given* $A \in M$ *the right coset* ΓA *contains a unique representative of the form*

(1) $$\begin{pmatrix} a & b \\ 0 & d \end{pmatrix}, \quad a,d \in \mathbb{N}, \quad 0 \le b < d.$$

Proof. In order to prove the existence apply (1.1) and assume $A = \begin{pmatrix} a & b \\ 0 & d \end{pmatrix}$, where $a \in \mathbb{N}$ and $d \in \mathbb{N}$ follows from $\det A = ad > 0$. Now choose $x \in \mathbb{Z}$ such that $0 \le b + xd < d$ and calculate

$$\begin{pmatrix} 1 & x \\ 0 & 1 \end{pmatrix}\begin{pmatrix} a & b \\ 0 & d \end{pmatrix} = \begin{pmatrix} a & b+xd \\ 0 & d \end{pmatrix}.$$

Let two matrices

$$\begin{pmatrix} a & b \\ 0 & d \end{pmatrix}, \quad \begin{pmatrix} \alpha & \beta \\ 0 & \delta \end{pmatrix}$$

of the type (1) be given. The calculation

$$\begin{pmatrix} a & b \\ 0 & d \end{pmatrix}\begin{pmatrix} \alpha & \beta \\ 0 & \delta \end{pmatrix}^{-1} = \begin{pmatrix} a/\alpha & (\alpha b - a\beta)/\alpha\delta \\ 0 & d/\delta \end{pmatrix}$$

shows that both matrices belong to the same right coset relative to Γ if and only if

$$a = \alpha, \quad d = \delta, \quad \text{and} \quad b = \beta. \qquad \square$$

The proposition immediately leads to

(1.3) Corollary. *Given* $\ell \in \mathbb{N}$ *the set* $M(\ell)$ *decomposes into*

$$\sigma_1(\ell) := \sum_{d \in \mathbb{N},\, d \mid \ell} d$$

right cosets relative to Γ. *A set of representatives is given by*

$$\begin{pmatrix} a & b \\ 0 & d \end{pmatrix}, \quad \textit{where } d \in \mathbb{N},\ d \mid \ell,\ 0 \le b < d \textit{ and } a = \ell/d.$$

In particular $(\Gamma, GL^+(2;\mathbb{Q}))$ *is a Hecke pair.*

Proof. The first part is clear according to (1.2). Given $A \in GL^+(2;\mathbb{Q})$ choose $\alpha \in \mathbb{N}$ such that $\alpha A \in M$. The assertion follows from

$$\#(\Gamma \backslash \Gamma A \Gamma) = \#(\Gamma \backslash \Gamma \alpha A \Gamma). \qquad \square$$

It is not necessary to choose the c-entry equal to 0 . Analogous arguments yield

(1.4) Proposition. *Given* $A \in M$ *the right coset* ΓA *contains a unique representative of the form*

$$\begin{pmatrix} a & 0 \\ c & d \end{pmatrix}, \quad a,d \in \mathbb{N}, \quad 0 \le c < a.$$

In the Exercises we develop a slightly more general example on the same explicit level. Therefore we fix some notations for the whole chapter. Let R denote a principal ideal domain (p.i.d.) with quotient field K (cf. [26]) . We write $\mathcal{U}$ for the group of units in R and call two elements $a,b \in R$ *associate*, if one has

$$a = \varepsilon b \quad \text{for some} \quad \varepsilon \in \mathcal{U}.$$

We call R *norm-finite*, if every non-zero ideal has finite index in R , i.e.,

$$N(a) := \#(R/aR) < \infty \quad \text{for all} \quad 0 \neq a \in R.$$

In this case we define $N(0) := 0$. Moreover we set

$$\mathrm{Inv}(n;R) := \mathrm{Mat}(n;R) \cap GL(n;K) = \{A \in \mathrm{Mat}(n;R) \mid \det A \neq 0\}.$$

(1.5) Exercises. Let R be a p.i.d. .

a) Verify the following equality

$$\begin{aligned} GL(2;R) &:= \{A \in \mathrm{Mat}(2;R) \mid \det A \neq 0,\ A^{-1} \in \mathrm{Mat}(2;R)\} \\ &= \{A \in \mathrm{Mat}(2;R) \mid \det A \in \mathcal{U}\}. \end{aligned}$$

b) Give a description of the right cosets $GL(2;R)A$, $A \in \mathrm{Inv}(2;R)$, in analogy with (1.2) .

c) Show that $(GL(2;R),GL(2;K))$ is a Hecke pair if and only if R is norm-finite.

d) Show that examples of norm-finite p.i.d.'s are given by an arbitrary field F , $\mathbb{Z}$, $\mathbb{Z}[i]$, $\mathbb{Z}[i\sqrt{2}]$, $\mathbb{Z}[(1+i\sqrt{3})/2]$, $\mathbb{Z}[(1+i\sqrt{7})/2]$, $\mathbb{Z}[(1+i\sqrt{11})/2]$, the polynomial ring $K[X]$, where $K = GF(q)$ is a finite field. Describe $N(a)$ in these cases.

e) Let R be a norm-finite p.i.d. . Show that

$$N(ab) = N(a)\,N(b)$$

holds for all $a,b \in R$. Demonstrate

$$\mathcal{U} = \{a \in R \mid N(a) = 1\}.$$

Note that for a given norm-finite p.i.d. R the pair (R,N) in general fails to be a Euclidean domain. $\mathbb{Z}[(1+i\sqrt{19})/2]$ is an example of the order in an imaginary quadratic number field with class number 1 , which fails to be Euclidean.

2. Double Coset Decompositions

We describe each double coset by certain invariants. The first obvious invariant is the determinant. In order to obtain another invariant set

$$\delta(A) := \text{g.c.d. of the entries of } A ,$$

whenever A is a non-zero integral matrix.

(2.1) Lemma a) $\delta(A)\,\delta(B) \mid \delta(AB)$

holds for all $A,B \in M$.

b) *Given* $A \in M$ *and* $U,V \in \Gamma$ *one has*

$$\delta(UAV) = \delta(A) .$$

Proof. a) $\frac{1}{\delta(A)\,\delta(B)}(AB) = \left(\frac{1}{\delta(A)}A\right)\left(\frac{1}{\delta(B)}B\right)$

proves to be integral.

b) Part a) especially yields

$$\delta(A) \mid \delta(UAV) .$$

The same conclusion applied to UAV instead of A, to U^{-1} instead of U, and to V^{-1} instead of V yields

$$\delta(UAV) \mid \delta(A) ,$$

hence the equality. □

We need another well-known number theoretical assertion.

(2.2) Lemma. *Let* $a,c,d \in \mathbb{Z}$ *such that* $a \neq 0$ *and* $\text{g.c.d.}(a,c,d) = 1$. *Then there exists an integer* $x \in \mathbb{Z}$ *satisfying*

$$\text{g.c.d.}(a,c + xd) = 1 .$$

Proof. Choose $x = \text{g.c.d.}(c^{|a|}, a)^{-1} a$. Suppose that p is a prime number dividing a as well as $c + xd$. If $p \mid c$ then the definition of x yields $p \nmid x$. Hence $p \mid d$ becomes a contradiction. If $p \nmid c$ then $p \mid x$ yields another contradiction. □

Now we prove the Elementary Divisor Theorem for 2×2 matrices.

(2.3) Theorem. *Given* $A \in M$ *there exist* $U,V \in \Gamma$ *and* $a,d \in \mathbb{N}$ *such that*

(1) $$UAV = \begin{pmatrix} a & 0 \\ 0 & d \end{pmatrix} , \quad a \mid d .$$

a *and* d *are uniquely determined by* A *and given by*

$$a = \delta(A)\ ,\quad d = (\det A)/\delta(A)\ .$$

Proof. The uniqueness of the entries a,d in (1) follows from (2.1) . In order to prove the existence we may assume $\delta(A) = 1$, since A can otherwise be replaced by $\frac{1}{\delta(A)}A$. In view of (1.4) we may already suppose that A has the form

$$\begin{pmatrix} a & 0 \\ c & d \end{pmatrix},\ a > 0\ ,\ d > 0,\quad \text{g.c.d.}\,(a,c,d) = \delta(A) = 1\ .$$

Next apply (2.2) and determine $x \in \mathbb{Z}$ with g.c.d. $(a, c+xd) = 1$. The entries of the first columm of

$$\bar{A} = \begin{pmatrix} a & 0 \\ c & d \end{pmatrix}\begin{pmatrix} 1 & 0 \\ x & 1 \end{pmatrix} = \begin{pmatrix} a & 0 \\ c+xd & d \end{pmatrix}$$

are relatively prime. Due to (1.1) there exists $U \in \Gamma$ such that

$$U\bar{A} = \begin{pmatrix} 1 & \bar{b} \\ 0 & ad \end{pmatrix}.$$

Then multiplication by $\begin{pmatrix} 1 & -\bar{b} \\ 0 & 1 \end{pmatrix}$ from the right hand side yields the desired form. □

Note that the above proof gives the matrices $U,V \in \Gamma$ explicitly. We can derive a result on the multiplicativity of the function δ .

(2.4) Theorem. *Let* $A \in \mathcal{M}(k)$ *and* $B \in \mathcal{M}(\ell)$, *where* $k,\ell \in \mathbb{N}$ *are relatively prime. Then one has*

$$\delta(AB) = \delta(A)\,\delta(B)\ .$$

Proof. First of all we may replace A by $\frac{1}{\delta(A)}A$ and B by $\frac{1}{\delta(B)}B$ in order to assume $\delta(A) = \delta(B) = 1$. According to (2.1) and (2.3) it suffices to show that one has for all $U = \begin{pmatrix} a & b \\ c & d \end{pmatrix} \in \Gamma$

$$1 = \delta\left(\begin{pmatrix} 1 & 0 \\ 0 & k \end{pmatrix} U \begin{pmatrix} 1 & 0 \\ 0 & \ell \end{pmatrix}\right) = \delta\left(\begin{pmatrix} a & \ell b \\ kc & k\ell d \end{pmatrix}\right).$$

Suppose that p is a prime dividing the entries of this matrix. Hence $p \mid a$ and $p \mid kc$. But $p \mid c$ would yield a contradiction to $\det U = 1$. Thus $p \mid k$ follows. Then $p \mid \ell b$ and g.c.d. $(k,\ell) = 1$ lead to $p \mid b$, which contradicts $\det U = 1$ again. □

One cannot omit the condition that k,ℓ are relatively prime.

(2.5) Example. Let $k \in \mathbb{N}$ and

$$A = \begin{pmatrix} 1 & 0 \\ 0 & k \end{pmatrix},\ B = \begin{pmatrix} k & 0 \\ 0 & 1 \end{pmatrix},\ AB = \begin{pmatrix} k & 0 \\ 0 & k \end{pmatrix},\ \delta(A) = \delta(B) = 1\ ,\ \delta(AB) = k\ .$$

(2.6) Exercises. a) Compute $U,V \in \Gamma$ such that UAV has the form (1) for the following matrices A :

$$\begin{pmatrix} 7 & 16 \\ 11 & 14 \end{pmatrix}, \quad \begin{pmatrix} 6 & 3 \\ -9 & 12 \end{pmatrix}, \quad \begin{pmatrix} 8 & 4 \\ 4 & 10 \end{pmatrix}.$$

b) Show that the modular group Γ is generated by the matrices

$$\begin{pmatrix} 1 & 1 \\ 0 & 1 \end{pmatrix} \text{ and } \begin{pmatrix} 0 & 1 \\ -1 & 0 \end{pmatrix}.$$

c) Find an example of matrices $A,B \in \mathcal{M}$ satisfying

$$\delta(AB) \neq \delta(BA).$$

d) Let $r,s,t \in \mathbb{C}$ with $\mathrm{Re}(s) > 1$, $\mathrm{Re}(2s+r) > 1$, $\mathrm{Re}(t) > 2$. Show that

$$\sum_{A:\Gamma\backslash \mathcal{M}/\Gamma} \delta(A)^{-r}(\det A)^{-s} = \zeta(s)\zeta(2s+r), \qquad \sum_{A:\Gamma\backslash \mathcal{M}} (\det A)^{-t} = \zeta(t)\zeta(t-1),$$

where ζ denotes the Riemann zeta-function.

e) Let R be a p.i.d.. Show that $GL(2;R)AGL(2;R)$, $A \in \mathrm{Inv}(2;R)$, contains a representative of the form

$$\begin{pmatrix} a & 0 \\ 0 & d \end{pmatrix}, \quad a,d \in R, \quad a \mid d,$$

where a and d are unique up to associates. Verify the analogue of (2.4).

f) Let R be a norm-finite p.i.d.. Show that the numbers of left and right cosets in every double coset $GL(2;R)AGL(2;R)$, $A \in \mathrm{Inv}(2;R)$, coincide.

g) Give an example of an integral domain R, where the analogue of (2.3) resp. of part e) does not come true.

3. The Hecke Algebra Associated with $(\Gamma,\mathcal{M})$

We investigate the Hecke algebra $\mathcal{H} = \mathcal{H}(\Gamma,\mathcal{M})$ over $R = \mathbb{Z}$.

(3.1) Lemma. a) *$\mathcal{H}$ is a commutative algebra.*

b) *Given $A \in \mathcal{M}$ and $0 \neq \lambda \in \mathbb{Z}$ one has*

$$(\Gamma A\Gamma)\cdot(\Gamma(\lambda I)\Gamma) = \Gamma(\lambda A)\Gamma.$$

c) *Given $A \in \mathcal{M}(k)$ and $B \in \mathcal{M}(\ell)$, where $k,\ell \in \mathbb{N}$ are relatively prime, one has*

$$(\Gamma A\Gamma)\cdot(\Gamma B\Gamma) = \Gamma(AB)\Gamma.$$

Proof. a) Transposing becomes an anti-automorphism of the Hecke pair $(\Gamma, GL^{+}(2;\mathbb{Q}))$ satisfying $\Gamma A\Gamma = \Gamma A'\Gamma$ by virtue of (2.3). Now apply I(7.2).

b) Use I(4.6).

c) (2.4) yields $\Gamma A \Gamma B \Gamma = \Gamma A B \Gamma$, hence

$$(\Gamma A \Gamma) \cdot (\Gamma B \Gamma) = t\ \Gamma(AB)\Gamma$$

for some $t \in \mathbb{N}$. Due to b) we may assume

$$A = \begin{pmatrix} 1 & 0 \\ 0 & k \end{pmatrix}, \quad B = \begin{pmatrix} 1 & 0 \\ 0 & \ell \end{pmatrix}.$$

Now choose representatives $\bar{B} = \begin{pmatrix} b_1 & b_2 \\ 0 & b_4 \end{pmatrix}$ of $\Gamma \backslash \Gamma B \Gamma$ in the form (1.2) . We compute

$$\begin{pmatrix} 1 & 0 \\ 0 & k\ell \end{pmatrix} \begin{pmatrix} b_1 & b_2 \\ 0 & b_4 \end{pmatrix}^{-1} = \begin{pmatrix} b_4/\ell & -b_2/\ell \\ 0 & kb_1 \end{pmatrix}.$$

This matrix is integral if and only if $\bar{B} = B$. Thus $t = 1$ follows. □

We want to decompose the Hecke algebra into subalgebras with simpler structures. Given a prime p the set

$$\mathcal{M}_p := \bigcup_{\ell=0}^{\infty} \mathcal{M}(p^\ell)$$

proves to be a submonoid of $\mathcal{M}$. Thus

$$\mathcal{H}_p := \mathcal{H}(\Gamma, \mathcal{M}_p)$$

becomes a subalgebra of $\mathcal{H}$. We call $\mathcal{H}_p$ the *p-primary component of* $\mathcal{H}$.

(3.2) Theorem. *The Hecke algebra $\mathcal{H}$ coincides with the tensor product of its primary components, i.e.,*

$$\mathcal{H} = \bigotimes_{p \text{ prime}} \mathcal{H}_p .$$

Proof. A basis of $\mathcal{H}$ resp. $\mathcal{H}_p$ is given by the double cosets. Each double coset possesses a unique representative in the form (2.3) . Each such matrix can be represented as a product of such matrices, whose entries are powers of a fixed prime. Now (3.1) completes the proof. □

Now it is convenient to introduce the following abbreviations

$$T(a,d) = \Gamma\begin{pmatrix} a & 0 \\ 0 & d \end{pmatrix}\Gamma \quad \text{for } a,d \in \mathbb{N},$$

$$T(\ell) = \sum_{A:\Gamma\backslash\mathcal{M}(\ell)/\Gamma} \Gamma A \Gamma .$$

For a prime p one especially has

$$T(1,p) = T(p) = \mathcal{M}(p) .$$

In this notation (1.3) , (2.3) , and (3.1) are reformulated as

(3.3) Lemma. *Let* $a, d, \ell \in \mathbb{N}$, *whenever they occur.*

a) $$T(\ell) = \sum_{ad=\ell, a|d} T(a,d) .$$

b) $$T(a,d) \cdot T(\ell,\ell) = T(\ell a, \ell d) .$$

c) $$\operatorname{ind} T(\ell) = \sigma_1(\ell) .$$

(3.4) Exercises. Let R be a norm-finite p.i.d. , which fails to be a field. Set

$$M(a) := \{ A \in \mathrm{Mat}(2;R) \mid \det A \in \mathcal{U}a \} \quad \text{for} \quad 0 \neq a \in R .$$

Then take the analogous definitions for $\mathcal{H} := \mathcal{H}(GL(2;R), \mathrm{Inv}(2;R))$ as in this section in the particular case $R = \mathbb{Z}$.

a) Verify the analogue of (3.1) for R in place of $\mathbb{Z}$.

b) Show that

$$\mathcal{H} = \bigotimes_{\pi} \mathcal{H}_{\pi}$$

holds, where π runs through a set of non-associate prime elements in R .

c) Show that

$$\operatorname{ind} T(a) = \#\big(GL(2;R) \backslash M(a)\big) = \sum_{d} N(d)$$

holds for $0 \neq a \in R$, where d runs through a set of non-associate divisors of a in R .

4. The Primary Component $\mathcal{H}_p$

In this section we describe the structure of the subalgebra $\mathcal{H}_p$.

(4.1) Theorem. *Given a prime* p *one has:*

a) $$\operatorname{ind} T(1,p^k) = \operatorname{ind} T(p^\ell, p^{k+\ell}) = p^{k-1}(p+1) \quad \textit{for all } k,\ell \geq 1 .$$

b) $$T(1,p^k) = T(p^k) - T(p,p) \cdot T(p^{k-2}) \quad \textit{for } k \geq 2 .$$

c) $$T(p)^2 = T(1,p^2) + (p+1)\, T(p,p) .$$

d) $$T(p) \cdot T(p^k) = T(p^{k+1}) + p\, T(p,p) \cdot T(p^{k-1}) \quad \textit{for } k \geq 1 .$$

e) $$T(p) \cdot T(1,p^k) = T(1,p^{k+1}) + p\, T(p,p) \cdot T(1,p^{k-1}) \quad \textit{for } k \geq 2 .$$

f) $$T(p^r) \cdot T(p^s) = \sum_{\nu=0}^{\min(r,s)} p^\nu\, T(p,p)^\nu \cdot T(p^{r+s-2\nu}) \quad \textit{for } r,s \geq 0 .$$

g) $$T(k) \cdot T(\ell) = \sum_{d \mid \mathrm{g.c.d.}(k,\ell)} d\, T(d,d) \cdot T(k\ell/d^2) \quad \textit{for } k,\ell \geq 1 .$$

Proof. a) According to (2.3) and (1.3) a set of representatives of $\Gamma\backslash T(1,p^k)$ is given by

$$\begin{pmatrix}1 & b\\ 0 & p^k\end{pmatrix},\ 0 \le b < p^k,\ \begin{pmatrix}p^k & 0\\ 0 & 1\end{pmatrix},\ \begin{pmatrix}p^{k-\nu} & b\\ 0 & p^\nu\end{pmatrix},\ 0 \le b < p^\nu,\ p \nmid b,\ \nu = 1,\dots,k-1 .$$

Finally use $\operatorname{ind} T(p^\ell,p^\ell) = 1$ and (3.3) .

b) Apply (3.3) .

c) This is a particular case of d) by virtue of b) for $k = 2$.

d) We compute

$$\text{(1)}\qquad T(p^k)\cdot T(p) = \sum_{B:\Gamma\backslash \mathcal{M}(p^{k+1})/\Gamma} t(B)\,\Gamma B\Gamma.$$

In view of part a) a set of representatives of $\Gamma\backslash T(p)$ is given by the matrices

$$A_j = \begin{pmatrix}1 & j\\ 0 & p\end{pmatrix},\ j = 0,\dots,p-1,\ A_p = \begin{pmatrix}p & 0\\ 0 & 1\end{pmatrix}.$$

Given $B \in \mathcal{M}(p^{k+1})$ with $p \mid \delta(B)$ then $pA_j^{-1} \in \mathrm{Mat}(2;\mathbb{Z})$ leads to

$$BA_j^{-1} \in \mathrm{Mat}(2;\mathbb{Z}) .$$

Hence $t(B) = p+1$ follows. There exists exactly one double coset $\Gamma B\Gamma$ in $\mathcal{M}(p^{k+1})$ with $p \nmid \delta(B)$ and we choose the representative $B = \begin{pmatrix}1 & 0\\ 0 & p^{k+1}\end{pmatrix}$. Then $BA_j^{-1} \in \mathrm{Mat}(2;\mathbb{Z})$ for $j = 0,\dots,p$ holds if and only if $j = 0$. Thus $t(B) = 1$ follows. According to b) we obtain

$$\begin{aligned} T(p^k)\cdot T(p) &= T(1,p^{k+1}) + (p+1)\sum_{B:\Gamma\backslash \mathcal{M}(p^{k+1})/\Gamma,\ p\mid\delta(B)} \Gamma B\Gamma \\ &= T(p^{k+1}) + p\,T(p,p)\cdot T(p^{k-1}) . \end{aligned}$$

e) Apply b) and c) in the following calculation

$$\begin{aligned} T(p)\cdot T(1,p^2) &= T(p)\cdot T(p^2) - T(p)\cdot T(p,p) \\ &= T(p^3) + (p-1)\,T(p,p)\cdot T(p) = T(1,p^3) + p\,T(p,p)\cdot T(p) . \end{aligned}$$

If $k \ge 3$ the same arguments yield

$$\begin{aligned} T(p)\cdot T(1,p^k) &= T(p)\cdot T(p^k) - T(p)\cdot T(p,p)\cdot T(p^{k-2}) \\ &= T(p^{k+1}) + p\,T(p,p)\cdot T(p^{k-1}) - T(p,p)\cdot T(p^{k-1}) - p\,T(p,p)^2\cdot T(p^{k-3}) \\ &= T(1,p^{k+1}) + p\,T(p,p)\cdot T(1,p^{k-1}) . \end{aligned}$$

f) We use induction on r , where $r = 0$ is obvious and $r = 1$ follows from d) . Therefore let $2 \le r \le s$. Hence d) and the induction hypothesis yield

$$\begin{aligned} &T(p^r)\cdot T(p^s) \\ &= T(p)\cdot T(p^{r-1})\cdot T(p^s) - p\,T(p,p)\cdot T(p^{r-2})\cdot T(p^s) \end{aligned}$$

$$= \sum_{\nu=0}^{r-1} p^{\nu}\, T(p)\cdot T(p,p)^{\nu}\cdot T\left(p^{r+s-1-2\nu}\right) - p\sum_{\nu=0}^{r-2} p^{\nu}\, T(p,p)^{\nu+1}\cdot T\left(p^{r+s-2-2\nu}\right)$$

$$= \sum_{\nu=0}^{r-1} p^{\nu}\, T(p,p)^{\nu}\cdot T\left(p^{r+s-2\nu}\right) + \sum_{\nu=0}^{r-1} p^{\nu+1}\, T(p,p)^{\nu+1}\cdot T\left(p^{r+s-2-2\nu}\right)$$

$$- \sum_{\nu=0}^{r-2} p^{\nu+1}\, T(p,p)^{\nu+1}\cdot T\left(p^{r+s-2-2\nu}\right)$$

$$= \sum_{\nu=0}^{r} p^{\nu}\, T(p,p)^{\nu}\cdot T\left(p^{r+s-2\nu}\right) .$$

g) Combine f) with (3.2) and use $T(p,p)^{\nu} = T(p^{\nu},p^{\nu})$ for $\nu \geq 0$. □

The final result is given in

(4.2) Theorem. *Let* p *be a prime. Then*

$$\mathcal{H}_p = \mathbb{Z}[T(p),\, T(p,p)]$$

holds, where $T(p)$ *and* $T(p,p)$ *are algebraically independent.*

Proof. A simple induction using (4.1c) and e) yields

$$T(1,p^k) \in \mathbb{Z}[T(p),\, T(p,p)] .$$

Now (3.3) implies

$$\mathcal{H}_p = \mathbb{Z}[T(p),\, T(p,p)] .$$

(2.3) shows that there exist exactly $\left[\frac{\ell}{2}\right]+1$ double cosets relative to Γ in $M(p^{\ell})$, where $[x]$ denotes the greatest integer $\leq x$. On the other hand a monomial

$$T(p,p)^{\nu}\; T(p)^{\mu} , \quad \nu,\mu \geq 0 ,$$

becomes a linear combination of double cosets in $M(p^{\ell})$ if and only if $2\nu + \mu = \ell$. Hence there exist $\left[\frac{\ell}{2}\right]+1$ monomials with this condition. Since $T(p)$ and $T(p,p)$ generate $\mathcal{H}_p$, they prove to be algebraically independent. □

The description of $T(p^{\ell})$ as a polynomial in $T(p)$ and $T(p,p)$ in (4.1) is not explicit, but it is possible to express $T(p^{\ell})$ as a generalized Chebyshev polynomial in $T(p)$ and $T(p,p)$.

One can also state (4.1) as an identity between formal power series over $\mathcal{H}_p$.

(4.3) Theorem. *Let* p *be a prime and* X *an indeterminate. Then*

$$\sum_{\ell=0}^{\infty} T(p^{\ell})\, X^{\ell} = \left(1 - T(p)\,X + p\,T(p,p)\,X^2\right)^{-1}$$

holds formally in X .

Proof. We apply (4.1d) in the following calculation

$$\Big(\sum_{\ell=0}^{\infty} T(p^\ell)\, X^\ell\Big)\cdot\Big(1 - T(p)X + p\,T(p,p)X^2\Big)$$
$$= 1 + \sum_{\ell=2}^{\infty}\Big(T(p^\ell) - T(p)\cdot T(p^{\ell-1}) + p\,T(p,p)\cdot T(p^{\ell-2})\Big)X^\ell = 1 . \qquad \square$$

In all the applications to elliptic modular forms and Maass wave forms one makes decisive use of this formal identity.

Finally a combination of (3.2) and (4.3) yields the following formal Euler product.

(4.4) Corollary. *The identity*

$$\sum_{\ell=1}^{\infty} T(\ell)\ell^{-s} = \prod_{p\,\text{prime}} \Big(1 - T(p)p^{-s} + T(p,p)p^{1-2s}\Big)^{-1}$$

holds formally in s .

Passing from $\mathbb{Z}$ to any field of characteristic 0 it is also possible to describe another set of generators, which is used in the applications.

(4.5) Corollary. *The Hecke algebra* $\mathcal{H}(\Gamma, M)$ *over* $\mathbb{Q}$ *is generated by*

$$T(\ell)\ ,\ \ell \in \mathbb{N}\ .$$

Proof. We obtain

$$(*) \qquad T(p,p) = \frac{1}{p}\Big(T(p)^2 - T(p^2)\Big)$$

for any prime p from (4.1d) . Now apply (4.2) and (3.2) . $\square$

It is clear from (*) that the elements $T(\ell)$, $\ell \in \mathbb{N}$, over $\mathbb{Z}$ generate a proper subalgebra of the Hecke algebra.

The results for the Hecke algebra $\bar{\mathcal{H}} = \mathcal{H}(\Gamma, GL^+(2;\mathbb{Q}))$ follow easily from our considerations. Again we obtain

$$\bar{\mathcal{H}} = \bigotimes_{p\ \text{prime}} \bar{\mathcal{H}}_p\ ,$$

where

$$\bar{\mathcal{H}}_p = \mathbb{Z}[T(p),\ T(p,p),\ T(p,p)^{-1}]$$

and $T(p,p)^{-1} = T(\frac{1}{p},\frac{1}{p}) = \Gamma(\frac{1}{p}I)\Gamma$.

Now it is clear that the p-component of the Hecke algebra $\mathcal{H}(\Gamma, SL(2;\mathbb{Q}))$ over $\mathbb{Z}$ consists of all polynomials over $\mathbb{Z}$ in $T(p)^2T(p,p)^{-1}$. Hence (4.1c) immediately leads to the following

(4.6) Corollary. *The Hecke algebra* $\mathcal{H}(\Gamma, SL(2;\mathbb{Q}))$ *over* $R := \mathbb{Z}$ *consists of all polynomials over* $\mathbb{Z}$ *in the elements*

$$\Gamma\begin{pmatrix} p^{-1} & 0 \\ 0 & p \end{pmatrix}\Gamma\ , \quad p \ \textit{prime number},$$

which are algebraically independent.

(4.7) Exercises. a) Apply (4.4) in order to obtain

$$\sum_{\ell=1}^{\infty} \operatorname{ind} T(\ell)\, \ell^{-s} \ = \ \zeta(s)\,\zeta(s-1)$$

for all $s \in \mathbb{C}$ with $\operatorname{Re}(s) > 2$, where ζ denotes the Riemann zeta-function.

b) Given $0 \neq A \in Mat(2;\mathbb{Q})$ define the denominator of A by

$$\nu(A) \ := \min\{\lambda \in \mathbb{N} \mid \lambda A \in Mat(2;\mathbb{Z})\}\ .$$

Show that

$$\nu(UAV) = \nu(A) \quad \text{for} \quad U, V \in \Gamma\ , \ 0 \neq A \in Mat(2;\mathbb{Q})\ .$$

Given $s \in \mathbb{C}$ with $\operatorname{Re}(s) > 1$ show that

$$\sum_{A:\Gamma\backslash SL(2;\mathbb{Q})/\Gamma} \nu(A)^{-s} \ = \ \zeta(s)\ .$$

c) Consider the Hecke algebra $\mathcal{H}_p$ over $F = \mathbb{Z}/p\mathbb{Z}$. Show that the elements $T(p^{\nu})$, $\nu \in \mathbb{N}_0$, form a basis of the subalgebra $F[T(p)]$.

d) Let R be a norm-finite p.i.d. and π a prime element in R. Derive the analogue of (4.1), (4.2), and (4.3) for $\mathcal{H}_\pi$. Show that

$$\sum_{a} T(a)\, N(a)^{-s} \ = \ \prod_{\pi} \left(1 - T(\pi)\, N(\pi)^{-s} + T(\pi,\pi)\, N(\pi)^{1-2s}\right)^{-1}$$

holds formally in s, where a runs through a set of non-associate, non-zero elements in R and π runs through a set of non-associate prime elements in R.

5. An Embedding

In this section we consider a particular case of the embedding I(6.4). Therefore let

$$\Gamma \ = \ \left\{\pm\begin{pmatrix} 1 & n \\ 0 & 1 \end{pmatrix} \mid n \in \mathbb{Z}\right\} \ \subset \ \Gamma\ ,$$

$$M_\infty \ = \ \left\{\begin{pmatrix} a & b \\ 0 & d \end{pmatrix} \mid a,b,d \in \mathbb{Z},\ ad > 0\right\} \ \subset \ M\ .$$

Each right coset $\Gamma_\infty A$, $A \in M_\infty$, clearly contains a unique representative of the form (1.2). Moreover one easily checks that each double coset $\Gamma_\infty A \Gamma_\infty$, $A \in M_\infty$, contains a unique representative of the form

(1) $$\begin{pmatrix} a & b \\ 0 & d \end{pmatrix}, \quad a,d \in \mathbb{N}\ , \ 0 \le b < \text{g.c.d.}(a,d)\ .$$

Due to (1.2) the conditions

$$\Gamma \cap M_\infty = \Gamma_\infty \,, \quad M = \Gamma M_\infty$$

are fulfilled. We obtain the embedding

$$\nu : \mathcal{R}(\Gamma,M) \to \mathcal{R}(\Gamma_\infty,M_\infty)\,,$$
$$\sum_{A:\Gamma\backslash M} u(\Gamma A)\,\Gamma A \mapsto \sum_{A:\Gamma_\infty\backslash M_\infty} u(\Gamma A)\,\Gamma_\infty A\,.$$

According to I(6.4) we get an embedding

$$\nu : \mathcal{H}(\Gamma,M) \to \mathcal{H}(\Gamma_\infty, M_\infty)$$

of the algebras. But $\mathcal{H}(\Gamma_\infty, M_\infty)$ does not have the same structure as $\mathcal{H}(\Gamma,M)$. Following A.N. Andrianov [4] , section 3.4.1 , we obtain

(5.1) Lemma. $\mathcal{H}(\Gamma_\infty,M_\infty)$ *fails to be commutative and contains zero-divisors.*

Proof. Given an integer $a > 1$ we observe that

$$\Gamma_\infty\begin{pmatrix} a & 0\\ 0 & 1\end{pmatrix}\Gamma_\infty \,,\ \Gamma_\infty\begin{pmatrix} a & 0\\ 0 & a\end{pmatrix}\Gamma_\infty \,,\ \Gamma_\infty\begin{pmatrix} a & 1\\ 0 & a\end{pmatrix}\Gamma_\infty \,,\ \Gamma_\infty\begin{pmatrix} a^2 & 0\\ 0 & a\end{pmatrix}\Gamma_\infty \,,\ \Gamma_\infty\begin{pmatrix} a^2 & 1\\ 0 & a\end{pmatrix}\Gamma_\infty$$

are mutually distinct double cosets decomposing into a single right coset. Now we get

$$\Gamma_\infty\begin{pmatrix} a & 0\\ 0 & 1\end{pmatrix}\Gamma_\infty \cdot \left[\Gamma_\infty\begin{pmatrix} a & 0\\ 0 & a\end{pmatrix}\Gamma_\infty - \Gamma_\infty\begin{pmatrix} a & 1\\ 0 & a\end{pmatrix}\Gamma_\infty\right] = \Gamma_\infty\begin{pmatrix} a^2 & 0\\ 0 & a\end{pmatrix}\Gamma_\infty - \Gamma_\infty\begin{pmatrix} a^2 & a\\ 0 & a\end{pmatrix}\Gamma_\infty = 0\,,$$

$$\left[\Gamma_\infty\begin{pmatrix} a & 0\\ 0 & a\end{pmatrix}\Gamma_\infty - \Gamma_\infty\begin{pmatrix} a & 1\\ 0 & a\end{pmatrix}\Gamma_\infty\right]\cdot \Gamma_\infty\begin{pmatrix} a & 0\\ 0 & 1\end{pmatrix}\Gamma_\infty = \Gamma_\infty\begin{pmatrix} a^2 & 0\\ 0 & a\end{pmatrix}\Gamma_\infty - \Gamma_\infty\begin{pmatrix} a^2 & 1\\ 0 & a\end{pmatrix}\Gamma_\infty \neq 0\,.$$

□

Given a prime p we define

$$T_-(p) := \Gamma_\infty\begin{pmatrix} p & 0\\ 0 & 1\end{pmatrix}\Gamma_\infty \,,\quad T_+(p) := \Gamma_\infty\begin{pmatrix} 1 & 0\\ 0 & p\end{pmatrix}\Gamma_\infty \,.$$

Now a straightforward calculation yields

(5.2) Lemma. a) $T_-(p) + T_+(p) = \nu(T(p))$,

b) $T_-(p) \cdot T_+(p) = p\,\Gamma_\infty\begin{pmatrix} p & 0\\ 0 & p\end{pmatrix}\Gamma_\infty = p\,\nu(T(p,p))$,

c) $T_+(p) \cdot T_-(p) = \sum_{j=0}^{p-1} \Gamma_\infty\begin{pmatrix} p & j\\ 0 & p\end{pmatrix}\Gamma_\infty$.

The importance of the lemma is emphasized by the connection with (4.3) . Under the embedding ν the polynomial on the right hand side of (4.3) decomposes in the polynomial ring over $\mathcal{H}(\Gamma_\infty,M_\infty)$. (5.2 a) and (5.2 b) immediately imply

(5.3) Corollary. *If* p *is a prime, one has*

$$1 - \nu(T(p))X + p\,\nu(T(p,p))X^2 = (1 - T_-(p)X)\cdot(1 - T_+(p)X)\,.$$

(5.4) Exercises. a) Let $A,B \in \mathcal{M}_\infty$ with g.c.d. (det A, det B) = 1 . Show that

$$(\Gamma_\infty A\Gamma_\infty)\cdot(\Gamma_\infty B\Gamma_\infty) = \Gamma_\infty AB\Gamma_\infty .$$

b) Let p be a prime. Show that $\Gamma_\infty\begin{pmatrix} p & 0 \\ 0 & 1\end{pmatrix}\Gamma_\infty$ commutes with a double coset $\Gamma_\infty\begin{pmatrix} a & b \\ 0 & d\end{pmatrix}\Gamma_\infty$ in $\mathcal{H}(\Gamma_\infty,\mathcal{M}_\infty)$ if and only if

$$\text{g.c.d. } (a,d) = \text{g.c.d. } (pa,d) \quad \text{and} \quad b(p-1) \equiv 0 \mod \text{g.c.d. } (a,d) .$$

c) Show that the set

$$\left\{ \Gamma_\infty(\ell I)\Gamma_\infty \mid \ell \in \mathbb{N} \right\}$$

is a basis of the center of $\mathcal{H}(\Gamma_\infty,\mathcal{M}_\infty)$.

d) Derive the analogues of this section for a norm-finite p.i.d. R , which is not a field.

6. Congruence Subgroups

In this section we introduce congruence subgroups. Some results of the preceding sections are also valid for congruence subgroups. This will be specified in the sequel.

Given a positive integer q and $A,B \in \mathrm{Mat}(n;\mathbb{Z})$ we define the *congruence* mod q for matrices via

$$A \equiv B \mod q \iff \frac{1}{q}(A-B) \in \mathrm{Mat}(n;\mathbb{Z}) .$$

This equivalence relation is clearly compatible with the ring structure of $\mathrm{Mat}(n;\mathbb{Z})$. Then define

$$\begin{aligned}\Gamma(q) &:= \{ U \in \Gamma \mid U \equiv I \mod q \} \\ &= \left\{ U = \begin{pmatrix} a & b \\ c & d\end{pmatrix} \in \Gamma \mid a \equiv d \equiv 1 \mod q , \ b \equiv c \equiv 0 \mod q \right\} .\end{aligned}$$

Using the formula for U^{-1} in (1.1) , one immediately verifies

(6.1) Lemma. *If* q *is a positive integer, then* $\Gamma(q)$ *is a normal subgroup of* Γ .

We call $\Gamma(q)$ the *principal congruence subgroup of level* q . We will investigate this subgroup more precisely. Therefore we need

(6.2) Lemma. *Let* $q \in \mathbb{N}$ *and* $A \in \mathrm{Mat}(2;\mathbb{Z})$ *with* $\det A \equiv 1 \mod q$. *Then* $B \in \mathrm{Mat}(2;\mathbb{Z})$ *exists such that*

$$\det (A+qB) = 1 .$$

Proof. Choose $U,V \in \Gamma$ and $a,d \in \mathbb{N}$ according to (2.3) such that

$$UAV = \begin{pmatrix} a & 0 \\ 0 & d\end{pmatrix} .$$

Due to the assumption we have

$$\det A = ad = 1 + rq \quad \text{for some } r \in \mathbb{Z} .$$

Hence we can choose

$$B = U^{-1}\begin{pmatrix} 0 & -r \\ r & -dr \end{pmatrix} V^{-1} \in \mathrm{Mat}(2;\mathbb{Z}) . \qquad \square$$

Application of (6.2) immediately yields

(6.3) Proposition. *Given a positive integer* q *the canonical projection*

$$\Pi : \mathrm{Mat}(2;\mathbb{Z}) \longrightarrow \mathrm{Mat}(2;\mathbb{Z}/q\mathbb{Z})$$

induces an isomorphism between the factor group $\Gamma/\Gamma(q)$ *and* $SL(2;\mathbb{Z}/q\mathbb{Z})$.

We can make use of (6.3) in order to calculate the index of $\Gamma(q)$ in Γ . As a convention p is always supposed to denote prime number.

(6.4) Proposition. *If* $q > 1$, *one has*

$$[\Gamma : \Gamma(q)] = \# SL(2;\mathbb{Z}/q\mathbb{Z}) = q^3 \prod_{p|q} (1-p^{-2}) .$$

Proof. (6.3) yields the first equality. If $q = p_1^{\nu_1} \cdot \ldots \cdot p_r^{\nu_r}$ with mutually distinct primes p_j , the Chinese Remainder Theorem yields an isomorphism

$$\mathbb{Z}/q\mathbb{Z} \cong \prod_{j=1}^{r} \mathbb{Z}/p_j^{\nu_j}\mathbb{Z} .$$

This gives an isomorphism

$$SL(2;\mathbb{Z}/q\mathbb{Z}) \cong \prod_{j=1}^{r} SL(2;\mathbb{Z}/p_j^{\nu_j}\mathbb{Z}) .$$

Let therefore $q = p^\nu$, $\nu \geq 1$, without restriction. The first columns of the matrices in $SL(2;\mathbb{Z}/q\mathbb{Z})$ are exactly the columns $\binom{a}{c} \in (\mathbb{Z}/q\mathbb{Z})^2$, where at least one entry fails to be a multiple of $p + q\mathbb{Z}$. Hence we have $p^{2\nu} - p^{2\nu-2}$ possibilities for the first column. Two matrices in $SL(2;\mathbb{Z}/q\mathbb{Z})$ with the same first column differ from each other by a factor $\begin{pmatrix} 1 & x \\ 0 & 1 \end{pmatrix}$, $x \in \mathbb{Z}/q\mathbb{Z}$, on the right hand side. Hence we have

$$\# SL(2;\mathbb{Z}/q\mathbb{Z}) = q^3(1-p^{-2}) \quad \text{for } q = p^\nu . \qquad \square$$

A subgroup S of the modular group Γ is called a *congruence subgroup*, if there exists $q \in \mathbb{N}$ satisfying $\Gamma(q) \subset S$. A congruence subgroup has finite index in the modular group Γ due to (6.4) .

The most important example of a congruence subgroup is

$$\Gamma_0(q) := \left\{ \begin{pmatrix} a & b \\ c & d \end{pmatrix} \in \Gamma \mid c \equiv 0 \mod q \right\} .$$

(6.5) Proposition. *Given* $q > 1$ *then* $\Gamma_0(q)$ *is a congruence subgroup, which fails to be normal in* Γ. *One has*

$$[\Gamma:\Gamma_0(q)] = q \prod_{p|q} (1+p^{-1}) .$$

Proof. Clearly $\Gamma_0(q)$ is a congruence subgroup, which fails to be normal in Γ due to

$$\begin{pmatrix} 0 & 1 \\ -1 & 0 \end{pmatrix} \begin{pmatrix} 1 & 1 \\ 0 & 1 \end{pmatrix} \begin{pmatrix} 0 & 1 \\ -1 & 0 \end{pmatrix}^{-1} = \begin{pmatrix} 1 & 0 \\ -1 & 1 \end{pmatrix} \notin \Gamma_0(q) .$$

Using the cononical projection Π we get

$$\Pi(\Gamma_0(q)) = \left\{ \begin{pmatrix} a & b \\ 0 & d \end{pmatrix} \in SL(2;\mathbb{Z}/q\mathbb{Z}) \right\}$$

Clearly $\Pi(\Gamma_0(q))$ has the order

$$q\,\varphi(q) = q^2 \prod_{p|q} (1-p^{-1}) ,$$

where φ is Euler's totient function. Then (6.4) implies

$$[\Gamma:\Gamma_0(q)] = \frac{[\Gamma:\Gamma(q)]}{[\Gamma_0(q):\Gamma(q)]} = q \prod_{p|q} (1+p^{-1}) . \qquad \square$$

The notion of a "principal congruence subgroup" is not standard in the literature. Some authors call

$$\bar{\Gamma}(q) = \{ U \in \Gamma \mid U \equiv \pm I \mod q \}$$

the principal congruence subgroup of level q. One has

$$-I \in \Gamma(q) \iff q = 2 .$$

Then (6.4) yields the formula for the index

$$[\Gamma:\bar{\Gamma}(q)] = \begin{cases} (1/2)\,q^3 \prod_{p|q} (1-p^{-2}) & \text{if } q > 2 , \\ 6 & \text{if } q = 2 . \end{cases}$$

(6.6) Exercises. a) Given $q \in \mathbb{N}$ show that

$$\Gamma_0(q) = A\Gamma A^{-1} \cap \Gamma , \quad \text{where } A = \begin{pmatrix} 0 & -1 \\ q & 0 \end{pmatrix} .$$

b) Let $\ell, m \in \mathbb{N}$ and set $q = \text{g.c.d.}(\ell, m)$. Show that

$$\Gamma(q) = \left\{ AB \mid A \in \Gamma(\ell) ,\ B \in \Gamma(m) \right\} .$$

c) Show that $\Gamma_0(2)$ is generated by the matrices

$$\begin{pmatrix} 1 & 0 \\ 2 & 1 \end{pmatrix} \text{ and } \begin{pmatrix} 1 & 1 \\ 0 & 1 \end{pmatrix} .$$

d) Given $q > 1$ show that

$$\Gamma_0^0(q) := \left\{ \begin{pmatrix} a & b \\ c & d \end{pmatrix} \in \Gamma \mid b \equiv c \equiv 0 \mod q \right\}$$

is a congruence subgroup. Compute its index in Γ. Determine all the values of q, for which $\Gamma_0^0(q)$ is normal in Γ.

e) Describe the Hecke algebra $\mathcal{H}(\Gamma_0(p),\Gamma)$, whenever p is a prime.

7. The Hecke Algebra Associated with $\Gamma_0(q)$

We fix an integer $q > 1$ and set $\Gamma_0 := \Gamma_0(q)$ throughout this section. $(\Gamma_0, GL^+(2;\mathbb{Q}))$ is a Hecke pair according to I(3.5c), (1.3), and (6.5). Following G. Shimura [58] we are interested in a particular subalgebra. Clearly

$$M_0 := \left\{ A = \begin{pmatrix} a & b \\ c & d \end{pmatrix} \in Mat(2;\mathbb{Z}) \mid \det A > 0 \, , \, c \equiv 0 \mod q \, , \, \text{g.c.d.}\,(a,q) = 1 \right\}$$

is a submonoid of $GL^+(2;\mathbb{Q})$ containing Γ_0.

(7.1) Lemma. *Given $A \in M_0$ the right coset $\Gamma_0 A$ contains a unique representative of the form*

(1) $$\begin{pmatrix} a & b \\ 0 & d \end{pmatrix}, \quad a,d \in \mathbb{N} \, , \, 0 \le b < d \, , \, \text{g.c.d.}\,(a,q) = 1 \, .$$

Proof. Apply (1.1) to the first column of A. Then $U \in \Gamma_0$ follows from the proof of (1.1). Now proceed as in (1.2). □

The proof of (2.3) must be slightly modified.

(7.2) Theorem. *Given $A \in M_0$ there exist $U,V \in \Gamma_0$ and $a,d \in \mathbb{N}$ such that*

(2) $$UAV = \begin{pmatrix} a & 0 \\ 0 & d \end{pmatrix}, \quad a \mid d \, , \, \text{g.c.d.}\,(a,q) = 1 \, .$$

a and d are uniquely determined and given by

$$a = \delta(A) \, , \quad d = (\det A)/\delta(A) \, .$$

Proof. In view of $\text{g.c.d.}\,(\delta(A),q) = 1$ we may suppose $\delta(A) = 1$ and due to (7.1)

$$A = \begin{pmatrix} a & b \\ 0 & d \end{pmatrix}, \quad a > 0 \, , \, d > 0 \, , \, \text{g.c.d.}\,(a,q) = \text{g.c.d.}\,(a,b,d) = 1$$

without restriction. Use (2.2) in order to find $x \in \mathbb{Z}$ with $\text{g.c.d.}\,(a,b+xd) = 1$ and set

$$\bar{A} = \begin{pmatrix} 1 & x \\ 0 & 1 \end{pmatrix} \begin{pmatrix} a & b \\ 0 & d \end{pmatrix} = \begin{pmatrix} a & b+xd \\ 0 & d \end{pmatrix} .$$

By virtue of $\text{g.c.d.}\,(a,q(b+xd)) = 1$ we can choose $\alpha,\beta \in \mathbb{Z}$ satisfying $\alpha a - \beta q(b+xd) = 1$, i.e.

$$U = \begin{pmatrix} a & b+xd \\ \beta q & \alpha \end{pmatrix} \in \Gamma_0 .$$

Then we obtain

$$\begin{pmatrix} 1 & 0 \\ \beta dq & 1 \end{pmatrix} \bar{A}\, U^{-1} = \begin{pmatrix} 1 & 0 \\ 0 & ad \end{pmatrix} . \qquad \square$$

We consider the Hecke algebra $\mathcal{H}_0 := \mathcal{H}(\Gamma_0, M_0)$ over $R = \mathbb{Z}$. Then exactly the same arguments as for (3.1) lead to

(7.3) Lemma. a) *$\mathcal{H}_0$ is a commutative algebra.*
b) *Given $A \in M_0$ and $\lambda \in \mathbb{Z}$ with g.c.d. $(\lambda, q) = 1$ one has*

$$(\Gamma_0 A \Gamma_0) \cdot (\Gamma_0 (\lambda I) \Gamma_0) = \Gamma_0 (\lambda A) \Gamma_0 .$$

c) *Let $A, B \in M_0$ with g.c.d. (det A, det B) = 1. Then one has*

$$(\Gamma_0 A \Gamma_0) \cdot (\Gamma_0 B \Gamma_0) = \Gamma_0 (AB) \Gamma_0 .$$

If p is a prime and $\ell \in \mathbb{N}$ we use the abbreviations

$$M_0(\ell) := \{ A \in M_0 \mid \det A = \ell \} ,$$
$$M_0^{(p)} := \bigcup_{\ell=0}^{\infty} M_0(p^\ell) ,$$
$$\mathcal{H}_0^{(p)} := \mathcal{H}(\Gamma_0, M_0^{(p)}) .$$

In analogy with (3.2) we get

(7.4) Corollary. *The Hecke algebra $\mathcal{H}_0$ coincides with the tensor product of its primary components, i.e.,*

$$\mathcal{H}_0 = \bigotimes_{p \text{ prime}} \mathcal{H}_0^{(p)} .$$

If $q = p_1^{\nu_1} \cdot \ldots \cdot p_r^{\nu_r}$ with $\nu_j \in \mathbb{N}$, we define for $\ell \in \mathbb{N}$

$$\ell \mid q^\infty \iff \ell = p_1^{\mu_1} \cdot \ldots \cdot p_r^{\mu_r} \quad \text{with } \mu_j \in \mathbb{N}_0 .$$

We can summarize the preceding results and relate the right coset decompositions for matrices in M_0 relative to Γ and to Γ_0.

(7.5) Lemma. a) *Let $A \in M_0(\ell)$ with g.c.d. $(\ell, q) = 1$. Then*

$$\Gamma A \Gamma = \bigcup \Gamma \begin{pmatrix} a & b \\ 0 & d \end{pmatrix} , \quad \Gamma_0 A \Gamma_0 = \bigcup \Gamma_0 \begin{pmatrix} a & b \\ 0 & d \end{pmatrix}$$

hold as disjoint unions, where

$$d \in \mathbb{N} ,\ d \mid \ell ,\ 0 \le b < d ,\ ad = \ell ,\ \text{g.c.d.}(a,b,d) = \delta(A) .$$

b) *If* $\ell \mid q^{\infty}$, *then* $M_0(\ell)$ *consists of the single double coset* $\Gamma_0\begin{pmatrix}1 & 0\\ 0 & \ell\end{pmatrix}\Gamma_0$ *and*

$$M_0(\ell) = \bigcup_{b=0}^{\ell-1} \Gamma_0\begin{pmatrix}1 & b\\ 0 & \ell\end{pmatrix}$$

holds as a disjoint union.

Proof. This is a consequence of (7.1) , (7.2) , (1.3) , and (2.3) . □

Now set

$$T_0(a,d) = \Gamma_0\begin{pmatrix}a & 0\\ 0 & d\end{pmatrix}\Gamma_0 \quad , \text{ if } a,d \in \mathbb{N} , \text{ g.c.d. } (a,q) = 1 ,$$

$$T_0(\ell) = \sum_{A:\Gamma_0\backslash M_0(\ell)/\Gamma_0} \Gamma_0 A \Gamma_0 .$$

Our main result is

(7.6) Theorem. a) *If* p *is a prime with* $p \nmid q$ *then* $\mathcal{H}_0^{(p)}$ *is the polynomial ring over* $\mathbb{Z}$ *in the elements*

$$T_0(p) , T_0(p,p) ,$$

which are algebraically independent. The mapping

$$\mathcal{H}_0^{(p)} \to \mathcal{H}_p ,$$
$$T_0(p) \mapsto T(p) , T_0(p,p) \mapsto T(p,p) ,$$

yields an isomorphism of the algebras.

b) *If* p *is a prime with* $p \mid q$, *then*

$$\mathcal{H}_0^{(p)} = \mathbb{Z}[T_0(p)]$$

holds, where $T_0(p)$ *is transcendent.*

c) *Given* $A \in M_0$ *one has a unique representation*

$$\Gamma_0 A \Gamma_0 = T_0(a,d) \cdot T_0(\ell) ,$$

where $a,d,\ell \in \mathbb{N}$, $a \mid d$, $\ell \mid q^{\infty}$, $a = \delta(A)$, *and* $ad\ell = \det A$.

d) $$T_0(\ell) \cdot T_0(m) = \sum_{\substack{d \mid \text{g.c.d.}(\ell,m)\\ \text{g.c.d.}(d,q)=1}} d\, T_0(d,d) \cdot T_0(\ell m/d^2)$$

holds for all $\ell,m \in \mathbb{N}$.

Proof. a) Combine (7.5) with I(6.4) and (4.2) .

b) (7.5) and the definition of the product yield

$$(*) \qquad T_0(p^{\ell}) = T_0(1,p^{\ell}) = T_0(p)^{\ell} \qquad \text{for all } \ell \in \mathbb{N} .$$

c) Apply (7.3) and b) .

d) Due to (7.4) we may assume $\ell = p^\nu$, $m = p^\mu$. If $p \mid q$ use (*) . In the case $p \nmid q$ the assertion follows from a) and (4.1g) . □

Again it is possible to express d) in terms of a formal power series resp. an Euler product (cf. (4.3) , (4.4)) .

(7.7) Corollary. a) *If* p *is a prime with* $p \mid q$, *then*

$$\sum_{\ell=0}^{\infty} T_0(p^\ell) X^\ell = (1 - T_0(p) X)^{-1}$$

holds formally in X .

b) *If* p *is a prime with* $p \nmid q$, *then*

$$\sum_{\ell=0}^{\infty} T_0(p^\ell) X^\ell = (1 - T_0(p) X + p\, T_0(p,p) X^2)^{-1}$$

holds formally in X .

c) $$\sum_{\ell=1}^{\infty} T_0(\ell)\, \ell^{-s} = \prod_{p \mid q} (1 - T_0(p) p^{-s})^{-1} \cdot \prod_{p \nmid q} (1 - T_0(p) p^{-s} + T_0(p,p) p^{1-2s})^{-1}$$

holds formally in s .

It is also clear by the same arguments as in (4.5) that the $T_0(\ell)$, $\ell \in \mathbb{N}$, generate the $\mathbb{Q}$-algebra $\mathcal{H}_0(\Gamma_0, M_0) \otimes_{\mathbb{Z}} \mathbb{Q}$.

(7.8) Exercises. a) Given $q > 1$ show that

$$\sum_{\ell=1}^{\infty} \operatorname{ind} T_0(\ell)\, \ell^{-s} = \Big(\prod_{p \mid q} (1 - p^{1-s}) \Big) \zeta(s) \zeta(s-1)$$

holds for all $s \in \mathbb{C}$ with $\operatorname{Re}(s) > 2$, where ζ denotes the Riemann zeta-function.

b) Given $q > 1$ define $W_q := \Gamma_0 \begin{pmatrix} 0 & -1 \\ q & 0 \end{pmatrix} \Gamma_0 \in \mathcal{H}(\Gamma_0, M)$. Show that W_q commutes with all the double cosets $\Gamma_0 A \Gamma_0$, $A \in M_0$, g.c.d.(det A,q) = 1 . Show that the Hecke algebra $\mathcal{H}(\Gamma_0, M)$ fails to be commutative.

c) Given $q > 1$ show that the mapping

$$\Gamma_0(q)\backslash\Gamma \longrightarrow \Gamma\backslash M(q) ,$$
$$\Gamma_0(q) A \longmapsto \Gamma \begin{pmatrix} 0 & -1 \\ q & 0 \end{pmatrix} A ,$$

is a bijection.

d) Let $q \in \mathbb{N}$ and $\Gamma_0^0 := \Gamma_0^0(q)$ as in (6.6) as well as

$$M_0^0 := \left\{ \begin{pmatrix} a & b \\ c & d \end{pmatrix} \in M_0 \mid b \equiv 0 \mod q \right\} .$$

Derive the analogue of (7.1) and (7.2) for right resp. double cosets relative to Γ_0^0 in M_0^0 . Are the Hecke algebras $\mathcal{H}(\Gamma_0^0, M_0^0)$ and $\mathcal{H}(\Gamma_0, M_0)$ isomorphic?

Remarks

This chapter refers to the roots of Hecke algebras. E. Hecke [20] discussed $\mathcal{H}(\Gamma,M)$, although he did not introduce the abstract Hecke algebra, but its linear representation in the space of elliptic modular forms. The endomorphisms of the space of elliptic modular forms, which are induced by $\mathcal{H}(\Gamma,M)$, are called *Hecke operators*. In fact L.J. Mordell [48] was the first, who studied Hecke operators in the particular case of the discriminant Δ. But it was E. Hecke's merit to recognize the universal meaning of this construction. The explicit version of the description of the Hecke algebra in (4.1) was given by G. Shimura [58], chapter III. There are several books on elliptic modular forms, which contain the description usually on the basis of Hecke operators, e.g., T.M. Apostol [6], R.C. Gunning [19], M. Koecher [33], H. Maaß [44], T. Miyake [47], A. Ogg [50], J.P. Serre [55], A. Terras [63]. Last not least there is of course the original work by E. Hecke [20]. Hecke operators for Maass wave forms are for instance investigated by H. Maaß [44] and A. Terras [63]. Moreover these results are also contained in the descriptions of the Hecke algebras related with the unimodular group and the Siegel modular group. References are contained in the next two chapters.

Chapter V

The Hecke Algebra Associated with the Unimodular Group

In this chapter we will generalize the results of the previous chapter to arbitrary degree. We will determine the structure of the Hecke algebra associated with the unimodular group of degree n. We include the number theoretical background material, e.g., Elementary Divisor Theorem or multiplicativity of elementary divisors. Moreover we derive T. Tamagawa's result on the rationality of a formal power series over this Hecke algebra.

$\mathrm{Mat}(n;\mathbb{Z})$ stands for the $\mathbb{Z}$-algebra of all integral $n\times n$ matrices. $\mathcal{U}_n := GL(n;\mathbb{Z})$ denotes the group of units in the ring $\mathrm{Mat}(n;\mathbb{Z})$, i.e.,

$$\mathcal{U}_n := GL(n;\mathbb{Z}) := \{U \in \mathrm{Mat}(n;\mathbb{Z}) \mid \text{there is } V \in \mathrm{Mat}(n;\mathbb{Z}) \text{ with } UV = VU = I\},$$

where I denotes the $n\times n$ identity matrix. $\mathcal{U}_n$ is called the *unimodular group of degree* n and its elements are said to be *unimodular matrices.*

The unimodular group is contained in the general linear group of degree n over $\mathbb{Q}$, i.e., in

$$GL(n;\mathbb{Q}) := \{A \in \mathrm{Mat}(n;\mathbb{Q}) \mid \det A \neq 0\}.$$

We will show that $(\mathcal{U}_n, GL(n;\mathbb{Q}))$ is a Hecke pair. For the sake of convenience we introduce the monoid

$$\mathrm{Inv}(n;\mathbb{Z}) := \{A \in \mathrm{Mat}(n;\mathbb{Z}) \mid \det A \neq 0\} = \mathrm{Mat}(n;\mathbb{Z}) \cap GL(n;\mathbb{Q})$$

of all (over $\mathbb{Q}$) invertible integral $n\times n$ matrices. We are going to describe the structure of the Hecke algebra

$$\mathcal{H}_n := \mathcal{H}(\mathcal{U}_n, \mathrm{Inv}(n;\mathbb{Z})) \quad \text{over } R := \mathbb{Z},$$

which turns out to be a polynomial ring over $\mathbb{Z}$ in infinitely many algebraically independent indeterminates.

The case $n = 1$ is obvious. Considering $\mathbb{N}$ as a multiplicative semi-group, the Hecke algebra $\mathcal{H}_1$ coincides with the semi-group algebra of $\mathbb{N}$ over $\mathbb{Z}$, i.e.,

$$\mathcal{H}_1 = \mathbb{Z}[\mathbb{N}].$$

1. The Unimodular Group

At first we give a handsome description of the unimodular group.

(1.1) Theorem. *Given* $U \in \mathrm{Mat}(n;\mathbb{Z})$ *the following assertions are equivalent:*

(i) U *is unimodular.*

(ii) U' *is unimodular.*

(iii) $\det U = \pm 1$.

(iv) *The map* $\mathbb{Z}^n \to \mathbb{Z}^n$, $x \mapsto Ux$, *is bijective.*

(v) *The map* $\mathbb{Z}^n \to \mathbb{Z}^n$, $x \mapsto Ux$, *is surjective.*

Proof. (i) $\Leftrightarrow$ (ii), (i) $\Rightarrow$ (iii), (i) $\Rightarrow$ (iv) $\Rightarrow$ (v): These implications are obvious.
(v) $\Rightarrow$ (i): Let $e_1,\dots,e_n$ denote the canonical basis of $\mathbb{Z}^n$. In view of the surjectivity there exist $v_1,\dots,v_n \in \mathbb{Z}^n$ such that $Uv_i = e_i$. Then $V = (v_1,\dots,v_n) \in \mathrm{Mat}(n;\mathbb{Z})$ satisfies $UV = I$ and $V = U^{-1}$ follows.
(iii) $\Rightarrow$ (i): Clearly U belongs to $\mathrm{Inv}(n;\mathbb{Z})$. The entries of U^{-1} are quotients of certain subdeterminants of U over $\det U$. Hence U^{-1} is integral. □

Note that one cannot add the condition

"The map $\mathbb{Z}^n \to \mathbb{Z}^n$, $x \mapsto Ux$, is injective."

as it is possible if $\mathbb{Z}$ is replaced by a field. $U = 2I$ provides a counterexample.

The subgroup

$$\mathcal{SU}_n = SL(n;\mathbb{Z}) = \{U \in \mathrm{Mat}(n;\mathbb{Z}) \mid \det U = 1\}$$

of $\mathcal{U}_n$ is a normal subgroup of index 2 . We call $\mathcal{SU}_n$ the *special unimodular group of degree* n . Clearly $\mathcal{SU}_2 = \Gamma$ holds in the notation of the previous chapter. Moreover one has

$$\mathcal{U}_n = \mathcal{SU}_n \cup F\,\mathcal{SU}_n ,$$

where $F = \mathrm{diag}(-1,1,\dots,1)$.

Observe that the case $n = 1$ becomes trivial in view of

$$\mathcal{U}_1 = \{\pm 1\} .$$

We also consider a generalization of unimodular matrices. A matrix $A \in \mathrm{Mat}(m,n;\mathbb{Z})$ is called *primitive*, if a matrix $B \in \mathrm{Mat}(n,m;\mathbb{Z})$ exists such that $BA = I^{(n)}$ in the case $m \geq n$ resp. $AB = I^{(m)}$ in the case $m \leq n$. If $m = n$ clearly unimodular and primitive matrices coincide. We will give several characterizations of primitive matrices in the Exercises.

Moreover recall the definition of a norm-finite p.i.d. and the notations from section IV.1 as well as the congruence for integral matrices from section IV.6 .

(1.2) Exercises. a) Show that $A \in \mathrm{Mat}(m,n;\mathbb{Z})$, $m \le n$, is primitive if and only if the map

$\mathbb{Z}^n \to \mathbb{Z}^m$, $x \mapsto Ax$, is surjective.

b) Given $q \in \mathbb{N}$ show that

$$\mathcal{U}_n(q) := \{ U \in \mathcal{U}_n \mid U \equiv I \mod q \}$$

is a normal subgroup of $\mathcal{U}_n$ with finite index. Compute its index in $\mathcal{U}_n$ provided that q is a prime.

c) Let R be a p.i.d. . Verify the analogue of (1.1) for R in place of $\mathbb{Z}$.

d) Let R be a norm-finite p.i.d. . Show that

$$GL(n;R) = \{ U \in \mathrm{Mat}(n;R) \mid N(\det U) = 1 \} .$$

2. The Elementary Divisor Theorem

In this section we prove the Elementary Divisor Theorem, which is the surrogate for the Bruhat decomposition in section III.4 .

Again let I_{jk} denote the matrix units in $\mathrm{Mat}(n;\mathbb{Z})$ with a single 1 as (j,k)-entry and 0 's otherwise. The permutation matrices P_π , $\pi \in S_n$, were already introduced in section III.2 . We identify S_n with the group of permutation matrices in $\mathcal{U}_n$ and recall the identity

(1) $$P_\pi I_{jk} P_\pi^{-1} = I_{\pi(j),\pi(k)} \quad \text{for } \pi \in S_n .$$

Now we derive a result on the generators of the unimodular group and on normal forms in double cosets.

(2.1) Theorem. *If* $n > 1$ *the unimodular group* $\mathcal{U}_n$ *is generated by the matrices*

(2) $$I + I_{12} , \quad P_k := P_{\tau_{1k}} = I - I_{11} - I_{kk} + I_{1k} + I_{k1} , \quad k = 2,\dots,n .$$

(2.2) Theorem. *Let* $A \in \mathrm{Mat}(m,n;\mathbb{Z})$ *with* $\mathrm{rank}\, A = r > 0$. *Then there exist* $U \in \mathcal{U}_m$, $V \in \mathcal{U}_n$, *and* $d_1,\dots,d_r \in \mathbb{N}$ *such that*

$$UAV = \begin{pmatrix} D & 0 \\ 0 & 0 \end{pmatrix} , \quad D = \mathrm{diag}\,(d_1,\dots,d_r) , \quad d_1 \mid d_2 \mid \dots \mid d_r .$$

Proof of (2.1) and (2.2). Let $G(n)$ denote the subgroup of $\mathcal{U}_n$ generated by the matrices (2) and set $G(1) = \{\pm 1\}$. In view of $(I + I_{12})^\ell = I + \ell I_{12}$ for $\ell \in \mathbb{Z}$ and the fact that S_n is generated by the transpositions τ_{1k} , $k = 2,\dots,n$, we conclude from (1) that

(3) $$S_n \subset G(n), \quad I + \ell I_{jk} \in G(n) , \quad \text{whenever } j \ne k \text{ and } \ell \in \mathbb{Z} .$$

Multiplication by the matrices in $G(n)$ therefore in particular yields

(4) interchange of two rows (resp. columns),

(5) addition of ℓ -times one row (resp. column) to another one, where $\ell \in \mathbb{Z}$.

At first we prove

Claim 1. *If* $n > 1$ *and* $U \in G(n-1)$, *then* $\begin{pmatrix} 1 & 0 \\ 0 & U \end{pmatrix} \in G(n)$.

Proof. In view of (3) only the case $n = 2$ remains to be proved:

$$\begin{pmatrix} 1 & 0 \\ -1 & 1 \end{pmatrix}\begin{pmatrix} 1 & 1 \\ 0 & 1 \end{pmatrix}\begin{pmatrix} 0 & 1 \\ 1 & 0 \end{pmatrix}\begin{pmatrix} 1 & -1 \\ 0 & 1 \end{pmatrix} = \begin{pmatrix} 1 & 0 \\ 0 & -1 \end{pmatrix} .$$ □

The following statement is proved by induction on q :

$S(q)$: Given $A \in \mathrm{Mat}(m,n;\mathbb{Z})$ with rank $A = r > 0$, where $m+n \le q$, there exist $U \in G(m)$, $V \in G(n)$, $d_1,\dots,d_r \in \mathbb{N}$ such that

$$UAV = \begin{pmatrix} D & 0 \\ 0 & 0 \end{pmatrix}, \quad D = \mathrm{diag}\,(d_1,\dots,d_r) , \quad d_1 \mid d_2 \mid \dots \mid d_r .$$

The case $q = 2$ is obvious. Therefore let $m+n = q+1$ and consider the double coset

$$\mathcal{A} = G(m)\,A\,G(n) .$$

The set of positive integers b , for which $B = (b_{jk}) \in \mathcal{A}$ and a pair (j,k) exist such that $b = b_{jk} > 0$, is denoted by $\mathcal{B}$. The set $\mathcal{B}$ is non-empty due to $-I \in G(n)$. Let

$$c := \min \mathcal{B} .$$

Claim 2. *If* $B = (b_{jk}) \in \mathcal{A}$ *with* $b_{11} = c$ *then one has*

$$c \mid b_{1k} , \; k = 2,\dots,n , \quad c \mid b_{j1} , \; j = 2,\dots,m .$$

Proof. Given $2 \le k \le n$ choose $\ell \in \mathbb{Z}$ with $0 \le b_{1k} + \ell c < c$. If we add ℓ times the first columm to the k-th columm, we conclude

$$b_{1k} + \ell c \in \mathcal{B} \cup \{0\}$$

from (5) and hence $b_{1k} = -\ell c$ because of the choice of c . The same conclusion holds for the elements of the first row. □

Claim 3. *There exists a matrix*

$$\begin{pmatrix} c & 0 \\ 0 & C \end{pmatrix} \in \mathcal{A} , \quad C \in \mathrm{Mat}(n-1,m-1;\mathbb{Z}) .$$

Proof. Choose a matrix $B \in \mathcal{A}$ with c as an entry. Now assume $b_{11} = c$ due to (4) . Then Claim 2 and (5) lead to a matrix of the form

$$\begin{pmatrix} c & 0 \\ 0 & C \end{pmatrix} \in \mathcal{A} , \quad C \in \mathrm{Mat}(n-1,m-1;\mathbb{Z}) .$$ □

If $m = 1$ or $n = 1$ or rank $A = 1$, the assertion is proved. Otherwise the induction hypothesis yields $U_1 \in G(m-1)$, $V_1 \in G(n-1)$ such that $U_1 C V_1$ has the desired form. According to Claim 1 we have

$$\begin{pmatrix} D & 0 \\ 0 & 0 \end{pmatrix} \in \mathcal{A}, \quad D = \operatorname{diag}(c, d_2, \ldots, d_r), \quad d_2 | d_3 | \ldots | d_r .$$

Next observe that

$$(I + I_{12}) \begin{pmatrix} D & 0 \\ 0 & 0 \end{pmatrix} = \begin{pmatrix} c & d_2 & 0 & \ldots & 0 \\ * & * & * & \ldots & * \end{pmatrix} \in \mathcal{A} .$$

Claim 2 yields $c | d_2$.

Hence (2.2) is proved.

Given $U \in \mathcal{U}_n$ apply $S(2n)$. Calculation of the determinants shows that the arising diagonal matrix must be the identity matrix. Thus $U \in G(n)$ follows. □□

In particular we get a result on the generators of the modular group (cf. IV(2.6)).

(2.3) Corollary. *The modular group* $\Gamma = SL(2;\mathbb{Z})$ *is generated by the matrices*

$$J = \begin{pmatrix} 0 & 1 \\ -1 & 0 \end{pmatrix} \quad \textit{and} \quad Q = \begin{pmatrix} 1 & 1 \\ 0 & 1 \end{pmatrix} .$$

Proof. In view of (2.1) the unimodular group $\mathcal{U}_2$ is generated by P_2 and Q. Hence the elements in Γ are words in P_2 and Q^ℓ, $\ell \in \mathbb{Z}$, where P_2 occurs in an even number. But one has for all $\ell \in \mathbb{Z}$:

$$P_2 Q^\ell P_2 = \begin{pmatrix} 1 & 0 \\ \ell & 1 \end{pmatrix} = J Q^{-\ell} J^{-1} .$$ □

(2.4) Remarks. a) The normal form in (2.2) is called *elementary divisor normal form*. Some authors call it *Smith normal form* (cf. M. Newman [49], p. 26).

b) The number of generators of $\mathcal{U}_n$ in (2.1) can be reduced for $n > 3$. $\mathcal{U}_n$ is always generated by the matrices

$$I + I_{12}, \quad I - 2I_{11} \quad \text{and} \quad P = \begin{pmatrix} 0 & 1 & 0 & \cdots & \cdots & 0 \\ \vdots & 0 & \ddots & & & \vdots \\ \vdots & & \ddots & \ddots & & \vdots \\ \vdots & & & \ddots & \ddots & \vdots \\ & & & & \ddots & 0 \\ 0 & & & & & 1 \\ (-1)^{n-1} & 0 & \cdots & & \cdots & 0 \end{pmatrix}$$

(cf. M. Newman [49], VII.5). In fact the special linear group $\mathcal{SU}_n$ is generated by

$$I + I_{12} \quad \text{and} \quad P .$$

c) The modular group Γ is also generated by the matrices

$$\begin{pmatrix} 1 & 1 \\ 0 & 1 \end{pmatrix} \quad \text{and} \quad \begin{pmatrix} 1 & 0 \\ 1 & 1 \end{pmatrix} .$$

(2.5) Exercises. a) Consider the matrix $A = \begin{pmatrix} 2 & 3 & 4 & 5 \\ 6 & 7 & 8 & 9 \end{pmatrix}$. Determine $U \in \mathcal{U}_2$ and $V \in \mathcal{U}_4$ such that UAV has elementary divisor normal form.
b) Show that the unimodular group $\mathcal{U}_n$ is generated by the set $\mathcal{U}_n \cap \mathrm{Sym}(n;\mathbb{Z})$.
c) Let $m,n \in \mathbb{Z}$, $A \in \mathrm{Inv}(m;\mathbb{Z})$, $B \in \mathrm{Inv}(n;\mathbb{Z})$. Show that $\mathrm{Mat}(m,n;\mathbb{Z})/A\,\mathrm{Mat}(m,n;\mathbb{Z})$ resp. $\mathrm{Mat}(m,n;\mathbb{Z})/\mathrm{Mat}(m,n;\mathbb{Z})B$ are finite $\mathbb{Z}$-modules of the order $|\det A|^n$ resp. $|\det B|^m$.
d) Show that a matrix $U \in \mathrm{Mat}(n;\mathbb{Z})$ is unimodular if and only if

$$\left\{ x \in \mathbb{Q}^n \mid Ux \in \mathbb{Z}^n \right\} = \mathbb{Z}^n .$$

e) Let $q,n \in \mathbb{N}$ and $A \in \mathrm{Mat}(n;\mathbb{Z})$ with $\det A \equiv 1 \bmod q$. Show that there is a matrix $B \in \mathrm{Mat}(n;\mathbb{Z})$ satisfying

$$\det(A + qB) = 1.$$

f) Derive the analogue of (2.2) for a p.i.d. R in place of $\mathbb{Z}$. If $n > 2$, show that the group $GL(n;R)$ is generated by the matrices

$$\begin{pmatrix} U & 0 \\ 0 & I \end{pmatrix}, \; U \in GL(2;R) \; , \; I = I^{(n-2)} \; , \; P_k \; , \; 2 \le k \le n .$$

3. Elementary Divisors

In order to deal with an arbitrary degree n , we need more than the two invariants introduced in chapter IV in order to describe the double cosets relative to $\mathcal{U}_n$.

Given $A \in \mathrm{Mat}(m,n;\mathbb{Z})$ let $\Delta_k(A)$, $1 \le k \le \min(m,n)$, denote the ideal in $\mathbb{Z}$ generated by all the $k \times k$ subdeterminants of A . Hence there exists a unique non-negative integer $\delta_k(A) \in \mathbb{N}_0$ satisfying

$$\Delta_k(A) = \mathbb{Z}\,\delta_k(A) .$$

Clearly $\delta_k(A)$ equals the g.c.d. of all the $k \times k$ subdeterminants of A . One easily checks

(1) $\delta_1(A) =$ g.c.d. of the entries of A .

(2) $\delta_k(A) = 0 \iff \Delta_k = \{0\} \iff k > \mathrm{rank}\, A$.

(3) $\delta_k(A') = \delta_k(A)$.

(4) $\delta_n(A) = |\det A|$, whenever $A \in \mathrm{Mat}(n;\mathbb{Z})$.

At first we show that all the δ_k's are actually invariants.

(3.1) Lemma. *Given* $A \in \mathrm{Mat}(m,n;\mathbb{Z})$, $U \in \mathcal{U}_m$, $V \in \mathcal{U}_n$, *and* $1 \le k \le \min(m,n)$, *one has*

$$\delta_k(UAV) = \delta_k(A) .$$

Proof. Due to the definition of δ_k and (3) it suffices to demonstrate

(5) $\Delta_k(AV) = \Delta_k(A)$.

Moreover we may restrict ourselves to the generators V of $\mathcal{U}_n$ given in (2.1) . If V is a permutation matrix, then (5) is obvious. Therefore let $n > 1$ and $V = I + I_{12}$. Given $\mu = (\mu_1,\ldots,\mu_k)$, $\nu = (\nu_1,\ldots,\nu_k)$, where $1 \le \mu_1 < \ldots < \mu_k \le m$, $1 \le \nu_1 < \ldots < \nu_k \le n$, let $A_{\mu,\nu}$ resp. $A^*_{\mu,\nu}$ denote the corresponding $k \times k$ submatrix of A resp. of AV . One has

$$\det A^*_{\mu,\nu} = \det A_{\mu,\nu} , \quad \text{if } \nu_1 \neq 2 ,$$

i.e., if $\nu_1 = 1$ or 2 does not occur among $\nu_1,\ldots,\nu_k$. Furthermore one has

$$\det A^*_{\mu,\nu} = \det A_{\mu,\nu} + \det A_{\mu,\nu^*} , \quad \text{where } \nu^* = (1,\nu_2,\ldots,\nu_k) , \text{ if } \nu_1 = 2 .$$

This leads to $\Delta_k(AV) = \Delta_k(A)$. □

One can also give a different proof (cf. M. Newman [49]). Therefore one uses the fact that $k = 1$ is obvious. Then let $A^{[k]}$ denote the k-th compound of A , which is the $\binom{m}{k} \times \binom{n}{k}$ matrix with all the $k \times k$ subdeterminants as its entries (cf. A.C. Aitken [1]). Then $\delta_k(A) = \delta_1(A^{[k]})$ holds. If B is an $n \times p$ matrix and $p \ge k$, the Cauchy-Binet formula yields

$$(6) \qquad (AB)^{[k]} = A^{[k]} B^{[k]} .$$

This formula leads to

$$\delta_k(A)\, \delta_k(B) \mid \delta_k(AB) ,$$

which implies (3.1) .

M. Koecher [32] proved a divisibility of the converse type, namely

$$(7) \qquad \delta_k(AB) = q_k\, \delta_k(A)\, \delta_k(B) , \text{ where } q_k \mid \text{g.c.d.}\left(\frac{\det A}{\delta_k(A)\delta_{n-k}(A)} , \frac{\det B}{\delta_k(B)\delta_{n-k}(B)}\right),$$

whenever $A,B \in \mathrm{Inv}(n;\mathbb{Z})$, $0 < k < n$.

(3.2) Lemma. *If* $A \in \mathrm{Mat}(m,n;\mathbb{Z})$ *and* $1 < k \le \min(m,n)$, *one has*

$$\delta_{k-1}(A) \mid \delta_k(A) .$$

Proof. $\Delta_k(A) \subset \Delta_{k-1}(A)$ follows from the expansion of a $k \times k$ subdeterminant according to a row, which yields an integral linear combination of $(k-1) \times (k-1)$ subdeterminants. □

According to (3.2) we obtain positive integers

$$\varepsilon_1(A) := \delta_1(A) , \quad \varepsilon_k(A) := \frac{\delta_k(A)}{\delta_{k-1}(A)} , \quad 1 < k \le \text{rank } A ,$$

whenever $A \in \mathrm{Mat}(m,n;\mathbb{Z})$. The integer $\varepsilon_k(A)$ is called the *k-th elementary divisor of* A . Some authors call $\varepsilon_1(A),\ldots,\varepsilon_r(A)$ the *invariant factors* of A .

If $d_1,\ldots,d_r \in \mathbb{N}$, where $d_1 \mid d_2 \mid \ldots \mid d_r$ and $D = \mathrm{diag}(d_1,\ldots,d_r)$, one has

$$(8) \qquad \delta_k(D) = d_1 \cdot \ldots \cdot d_k , \quad \varepsilon_k(D) = d_k , \quad k = 1,\ldots,r .$$

Hence a combination of (2.2) , (8) , and (3.1) yields the uniqueness in the Elementary Divisor Theorem.

(3.3) Theorem. *Elementary Divisor Theorem*
Let $A \in \mathrm{Mat}(m,n;\mathbb{Z})$ *with* $\operatorname{rank} A = r > 0$. *Then the double coset* $\mathcal{U}_m A \mathcal{U}_n$ *contains a unique representative of the form*

$$\begin{pmatrix} D & 0 \\ 0 & 0 \end{pmatrix}, \quad D = \operatorname{diag}(d_1,\ldots,d_r), \quad d_1,\ldots,d_r \in \mathbb{N}, \quad d_1 \mid d_2 \mid \ldots \mid d_r ,$$

where

$$d_k = \varepsilon_k(A), \quad 1 \le k \le r .$$

We rewrite (3.3) for a columm vector as

(3.4) Corollary. *Given* $0 \neq g \in \mathbb{Z}^n$ *there exists* $U \in \mathcal{U}_n$ *such that*

$$Ug = (\delta,0,\ldots,0)', \quad \delta = \delta_1(g) = \text{g.c.d. } (g) .$$

An induction starting with the first resp. last column leads to

(3.5) Corollary. *Given* $A \in \mathrm{Mat}(n;\mathbb{Z})$ *there exist* $U,V \in \mathcal{U}_n$ *such that* UA *becomes an upper and* VA *a lower triangular matrix.*

(3.3) moreover implies

(3.6) Corollary. *Given* $A \in \mathrm{Mat}(m,n;\mathbb{Z})$ *with* $\operatorname{rank} A = r > 0$, *one has*

a) $\varepsilon_{k-1}(A) \mid \varepsilon_k(A)$ *for* $1 < k \le r$,

b) $\delta_k^2(A) \mid \delta_{k-1}(A)\delta_{k+1}(A)$ *for* $1 < k < \min(m,n)$.

Now we involve some analysis. Let ζ denote the Riemann zeta-function

$$\zeta(s) = \sum_{m=1}^{\infty} m^{-s} ,$$

which converges absolutely for $s \in \mathbb{C}$, $\operatorname{Re}(s) > 1$.
We can associate a zeta-function to the double coset space $\mathcal{H}(\mathcal{U}_n, \mathrm{Inv}(n;\mathbb{Z}))$.

(3.7) Lemma. *Given* $s \in \mathbb{C}$ *with* $\operatorname{Re}(s) > 1$ *one has*

$$\zeta_n(s) = \sum_{D:\mathcal{U}_n \backslash \mathrm{Inv}(n;\mathbb{Z})/\mathcal{U}_n} |\det D|^{-s} = \prod_{\ell=1}^{n} \zeta(\ell s) .$$

Proof. In view of (3.3) we have to extend the summation over

$$D = \operatorname{diag}(k_1, k_1k_2, \ldots, k_1 \cdot \ldots \cdot k_n), \quad k_j \in \mathbb{N}, \quad j = 1,\ldots,n .$$

This leads to

$$\zeta_n(s) = \sum_{k_1,\dots,k_n \in \mathbb{N}} \left(k_1^{\,n} k_2^{\,n-1} \cdot \ldots \cdot k_n\right)^{-s} = \prod_{\ell=1}^{n} \zeta(\ell s) . \qquad \square$$

(3.8) Exercises. a) Determine the elementary divisors of the following matrices.

$$\begin{pmatrix} -3 & 4 & 5 \\ -1 & 2 & -4 \end{pmatrix}, \quad \begin{pmatrix} 1 & 1 & 0 & 0 \\ 1 & 1 & 1 & 0 \\ 0 & 1 & 1 & 1 \\ 0 & 0 & 1 & 1 \end{pmatrix}, \quad \begin{pmatrix} 2 & 1 & 1 & 1 \\ 1 & 2 & 1 & 1 \\ 1 & 1 & 2 & 1 \\ 1 & 1 & 1 & 2 \end{pmatrix}.$$

b) Given $A \in \mathrm{Mat}(m,n;\mathbb{Z})$, $m \geq n$, show that the following assertions are equivalent:

(i) A is primitive.

(ii) A' is primitive.

(iii) UAV is primitive for $U \in \mathcal{U}_m$, $V \in \mathcal{U}_n$.

(iv) There exists $W = (A,*) \in \mathcal{U}_m$.

(v) $\delta_n(A) = 1$.

(vi) rank $A = n$ and $\varepsilon_n(A) = 1$.

(vii) $\{x \in \mathbb{Q}^n \mid Ax \in \mathbb{Z}^m\} = \mathbb{Z}^n$.

c) Let $A \in \mathrm{Inv}(n;\mathbb{Z})$ and $0 \neq b \in \mathbb{Z}$. Determine the elementary divisors of

$$C = \begin{pmatrix} A & 0 \\ 0 & b \end{pmatrix} \in \mathrm{Inv}(n+1;\mathbb{Z}) .$$

d) Given $A \in \mathrm{Inv}(n;\mathbb{Z})$ define

$$\rho(A) := \text{g.c.d.} \left\{ a \in \mathbb{N} \mid aA^{-1} \in \mathrm{Mat}(n;\mathbb{Z}) \right\} .$$

Show that $\rho(A) = \varepsilon_n(A)$ holds and that $\rho(A)A^{-1}$ belongs to $\mathrm{Mat}(n;\mathbb{Z})$.

e) Let $(s_1,\dots,s_n)' \in \mathbb{C}^n$. Show that

$$\sum_{D:\mathcal{U}_n \backslash \mathrm{Inv}(n;\mathbb{Z})/\mathcal{U}_n} \varepsilon_1(A)^{-s_1} \cdot \ldots \cdot \varepsilon_n(A)^{-s_n} = \prod_{j=1}^{n} \zeta\left(s_j + s_{j+1} + \ldots + s_n\right) ,$$

$$\sum_{D:\mathcal{U}_n \backslash \mathrm{Inv}(n;\mathbb{Z})/\mathcal{U}_n} \delta_1(A)^{-s_1} \cdot \ldots \cdot \delta_n(A)^{-s_n} = \prod_{j=1}^{n} \zeta\left(s_j + 2s_{j+1} + \ldots + (n+1-j)s_n\right) ,$$

if the real parts of all the arguments on the right hand sides are greater than 1 .

f) Verify the analogues of (3.1) , (3.2) , (3.3) for a Euclidean domain R along the same lines resp. for an arbitrary p.i.d. R by the use of (6) .

4. q-Kernels

In this section we give another invariant description for the double cosets relative to $\mathcal{U}_n$. This is a sort of "local-global-principle". Properties of a matrix modulo all prime powers determine the double coset uniquely.

Given $A \in \mathrm{Mat}(n;\mathbb{Z})$ and $q \in \mathbb{N}$ we define the *q-kernel* of A by

(1) $$\ker(A \mid q) := \{x \in \mathbb{Z}^n \bmod q \mid Ax \equiv 0 \bmod q\} .$$

(4.1) Lemma. *Given* $A \in \mathrm{Mat}(n;\mathbb{Z})$ *with* $\operatorname{rank} A = r > 0$ *and* $q \in \mathbb{N}$ *one has*

a) $\#\ker(UAV \mid q) = \#\ker(A \mid q)$ *for all* $U, V \in \mathcal{U}_n$,

b) $\#\ker(A \mid q) = q^{n-r} \prod_{k=1}^{r} \text{g.c.d.}\left(q, \varepsilon_k(A)\right)$.

Proof. a) This part follows from (1.1iv) .
b) Due to a) and (3.3) we may restrict ourselves to

$$A = \operatorname{diag}(d_1,\dots,d_n)\ ,\quad d_k = \varepsilon_k(A) \ \text{ for } 1 \le k \le r\ ,\quad d_k = 0 \ \text{ for } r < k \le n\ .$$

Thus it suffices to consider $n = 1$ and $a \in \mathbb{Z}$, where

$$\#\ker(a \mid q) = \#\{x \in \mathbb{Z} \bmod q \mid ax \equiv 0 \bmod q\} = \text{g.c.d.}\,(a,q)$$

yields the assertion. □

The lemma gives another characterization of unimodular matrices, which can be added to (1.1) and can be viewed as the missing injectivity of the mappings.

(4.2) Corollary. *A matrix* $U \in \mathrm{Mat}(n;\mathbb{Z})$ *is unimodular if and only if*

(2) $\#\ker(U \mid p) = 1$ *for all primes* p .

The condition (2) means that $p \nmid \det U$ in view of (4.1) . Another consequence is derived in

(4.3) Lemma. *Let* $A, B \in \mathrm{Mat}(n;\mathbb{Z})$, $q \in \mathbb{N}$ *such that* $\text{g.c.d.}\,(q, \det B) = 1$. *Then one has*

$$\#\ker(AB \mid q) = \#\ker(BA \mid q) = \#\ker(A \mid q)\ .$$

Proof. It follows from (4.1b) that Bx runs through a set of representatives of the columns $\mathbb{Z}^n \bmod q$ if and only if x does. Then the definition yields the assertion. □

Now we give a characterization of double cosets, which involves elementary divisors as well as properties mod q .

(4.4) Theorem. *Given* $A, B \in \mathrm{Mat}(n;\mathbb{Z})$ *the following assertions are equivalent:*

(i) $\mathcal{U}_n A \mathcal{U}_n = \mathcal{U}_n B \mathcal{U}_n$.

(ii) $\operatorname{rank} A = \operatorname{rank} B = r$ *and* $\varepsilon_k(A) = \varepsilon_k(B)$ *for* $1 \le k \le r$.

(iii) $\delta_k(A) = \delta_k(B)$ *for* $1 \le k \le n$.

(iv) $\#\ker(A \mid q) = \#\ker(B \mid q)$ *for all* $q \in \mathbb{N}$.

(v) $\#\ker\left(A \mid p^\ell\right) = \#\ker\left(B \mid p^\ell\right)$ *for all primes* p *and* $\ell \in \mathbb{N}$.

Proof. (i) $\Leftrightarrow$ (ii) $\Leftrightarrow$ (iii): Apply (3.3) .
(i) $\Rightarrow$ (iv): Use (4.1) .
(iv) $\Rightarrow$ (v): Obvious.
(v) $\Rightarrow$ (ii): Due to (4.1) and (3.3) we may assume

$$A = \mathrm{diag}(a_1,\dots,a_n)\ ,\quad a_1 \mid \dots \mid a_n\ ,\quad B = \mathrm{diag}(b_1,\dots,b_n)\ ,\quad b_1 \mid \dots \mid b_n\ ,$$

without restriction, where all the entries are non-negative. Assume $A \neq B$. Then choose k such that

$$a_i = b_i \text{ for } i < k \text{ and } a_k \neq b_k\ .$$

Now determine a prime p and $\ell \in \mathbb{N}$ such that

$$p^\ell \mid a_k \text{ and } p^\ell \nmid b_k \text{ resp. } p^\ell \nmid a_k \text{ and } p^\ell \mid b_k\ .$$

Due to symmetry we assume $p^\ell \mid a_k$, $p^\ell \nmid b_k$. Then (4.1) implies

$$\# \ker(A \mid p^\ell) = \Big(\prod_{i=1}^{k-1} \text{g.c.d.}\,(a_i,p^\ell)\Big)\ p^{\ell(n+1-k)}$$
$$> \Big(\prod_{i=1}^{k-1} \text{g.c.d.}\,(a_i,p^\ell)\Big)\Big(\prod_{i=k}^{n} \text{g.c.d.}\,(b_i,p^\ell)\Big) = \# \ker(B \mid p^\ell)\ .$$

This is a contradiction and $A = B$ follows. □

Given $A \in \mathrm{Inv}(n;\mathbb{Z})$ denote the elementary divisor matrix of A by

$$E(A) = \mathrm{diag}\big(\varepsilon_1(A),\dots,\varepsilon_n(A)\big)\ .$$

The multiplicativity is now derived in

(4.5) Theorem. *Let* $A,B \in \mathrm{Inv}(n;\mathbb{Z})$ *such that* $\det A$ *and* $\det B$ *are relatively prime. Then one has*

$$E(AB) = E(A)\,E(B)\ .$$

Proof. There exist $U,V,W \in \mathcal{U}_n$ such that

$$E(AB) = U\,E(A)\,V\,E(B)\,W\ .$$

Since $\det A$ and $\det B$ are relatively prime, (4.1) and (4.3) imply

$$\# \ker\big(E(AB) \mid p^\ell\big) = \# \ker\big(A \mid p^\ell\big)\ \# \ker\big(B \mid p^\ell\big) = \# \ker\big(E(A)\,E(B) \mid p^\ell\big)$$

for all primes p and $\ell \in \mathbb{N}$. Now (4.4) leads to

$$E(AB) = E(A)\,E(B)\ .$$ □

Of course (4.5) follows immediately from M. Koecher's result (7) in section 3 . An alternative proof is given by M. Newman [49] , Theorem II.15 .

(4.6) Exercises. a) Use the analogue of (1) also for rectangular matrices. Show that a matrix $A \in \mathrm{Mat}(m,n;\mathbb{Z})$, $m \geq n$, is primitive if and only if

$$\# \ker(A|p) = 1 \quad \text{for all primes } p .$$

b) Let $A \in \mathrm{Mat}(n;\mathbb{Z})$ and define the arithmetical function

$$k_A(q) := \# \ker(A|q) , \quad q \in \mathbb{N} .$$

Show that k_A is multiplicative, i.e.,

$$k_A(\ell m) = k_A(\ell)\, k_A(m) , \quad \text{whenever g.c.d.}(\ell,m) = 1 .$$

c) Let $A \in \mathrm{Inv}(n;\mathbb{Z})$, $q \in \mathbb{N}$. Show that $\# \ker(A|q)$ always divides $\det A$ and that $\# \ker(A|q) = |\det A|$ holds if and only if qA^{-1} is integral again.
d) Give an example of matrices $A,B \in \mathrm{Inv}(n;\mathbb{Z})$ such that $E(AB) \neq E(BA)$.
e) Derive the analogous results for a norm-finite p.i.d. R in place of $\mathbb{Z}$.

5. Right Coset Decompositions

Recall that $\mathrm{Inv}(n;\mathbb{Z}) = \mathrm{Mat}(n;\mathbb{Z}) \cap GL(n;\mathbb{Q})$. At first we are going to strengthen (3.5) .

(5.1) Theorem. *Given* $A \in \mathrm{Inv}(n;\mathbb{Z})$ *the right coset* $\mathcal{U}_n A$ *contains a unique representative* B *of the form*

$$B = \begin{pmatrix} b_1 & 0 & \cdots & \cdots & 0 \\ b_{2,1} & b_2 & \ddots & & \vdots \\ \vdots & & \ddots & \ddots & \vdots \\ \vdots & & & \ddots & 0 \\ b_{n,1} & \cdots & \cdots & b_{n,n-1} & b_n \end{pmatrix} ,$$

where

$$b_1,\ldots,b_n \in \mathbb{N} , \quad 0 \leq b_{ij} < b_j \quad \textit{for} \quad 1 \leq j < i \leq n .$$

Proof. Existence and uniqueness are proved by induction on n , where $n = 1$ is obvious. Therefore let $n > 1$ and assume that A is a lower triangular matrix due to (3.5) . Hence let

$$A = \begin{pmatrix} a_1 & 0 \\ a & A_1 \end{pmatrix} , \quad 0 \neq a_1 \in \mathbb{Z} , \quad a \in \mathbb{Z}^{n-1} , \quad A_1 \in \mathrm{Inv}(n-1;\mathbb{Z}) .$$

Now choose

(1) $$U = \begin{pmatrix} \varepsilon & 0 \\ u & U_1 \end{pmatrix} \in \mathcal{U}_n , \quad \varepsilon = \pm 1 , \; u \in \mathbb{Z}^{n-1} , \quad U_1 \in \mathcal{U}_{n-1} ,$$

and calculate

$$(2) \qquad UA = \begin{pmatrix} \varepsilon a_1 & 0 \\ U_1 a + a_1 u & U_1 A_1 \end{pmatrix} .$$

In view of the induction hypothesis we can choose $U_1 \in \mathcal{U}_{n-1}$ such that $U_1 A_1$ has the desired form. Determine ε such that $\varepsilon a_1 > 0$ and $u \in \mathbb{Z}^{n-1}$ such that the entries of the column $U_1 a + a_1 u$ belong to $\{0,1,\dots,|a_1|-1\}$. Hence the existence follows.

Now assume that A and B have the desired form and that $U \in \mathcal{U}_n$ exists such that $UA = B$. We note that $U = BA^{-1}$ proves to be a lower triangular matrix. Thus U has the form (1). If B is given by

$$B = \begin{pmatrix} b_1 & 0 \\ b & B_1 \end{pmatrix}, \ b_1 \in \mathbb{N}, \ b \in \mathbb{Z}^{n-1}, \ B_1 \in \mathrm{Inv}(n-1;\mathbb{Z}),$$

then (2) yields

$$\varepsilon a_1 = b_1 , \ U_1 A_1 = B_1 , \ U_1 a + a_1 u = b .$$

$a_1, b_1 \in \mathbb{N}$ leads to $\varepsilon = 1$, the induction hypothesis gives $U_1 = I$ and finally $a + a_1 u = b$ implies $u = 0$, thus $U = I$ and $A = B$. □

The analogous arguments can be applied in order to obtain a unique upper triangular matrix in each right coset.

(5.2) Corollary. *Given* $A \in \mathrm{Inv}(n;\mathbb{Z})$ *the right coset* $\mathcal{U}_n A$ *contains a unique representative* C *of the form*

$$C = \begin{pmatrix} c_1 & c_{1,2} & \cdot & \cdot & \cdot & \cdot & c_{1,n} \\ 0 & c_2 & \cdot & & & & \cdot \\ \cdot & \cdot & \cdot & \cdot & & & \cdot \\ \cdot & & \cdot & \cdot & \cdot & & \cdot \\ \cdot & & & \cdot & \cdot & & c_{n-1,n} \\ 0 & \cdot & \cdot & \cdot & 0 & & c_n \end{pmatrix},$$

where

$$c_1,\dots,c_n \in \mathbb{N}, \quad 0 \le c_{ij} < c_j \quad \textit{for} \quad 1 \le i < j \le n .$$

Now we generalize the result IV(1.3) on the number of right cosets in a double coset to arbitrary degree n. We do not obtain an explicit formula, but at least the finiteness of this number.

(5.3) Corollary. *Given* $\ell \in \mathbb{N}$ *the set*

$$\mathcal{D}_n(\ell) := \{ A \in \mathrm{Mat}(n;\mathbb{Z}) \mid |\det A| = \ell \}$$

decomposes into finitely many right cosets relative to $\mathcal{U}_n$. *In particular* $(\mathcal{U}_n, \mathrm{GL}(n;\mathbb{Q}))$ *is a Hecke pair.*

Proof. Choose a representative B of the form (5.1) in each right coset of $\mathcal{D}_n(\ell)$. There are only finitely many possibilities for the diagonal entries, since they are divisors of ℓ. Hence there are only finitely many B 's . □

From IV(1.3) we know that

$$\#\,(\mathcal{U}_2\backslash\mathcal{D}_2(\ell)) = \sigma_1(\ell) = \sum_{d\mid\ell} d\,.$$

Similarly (5.1) yields

$$(3)\qquad \#\,(\mathcal{U}_3\backslash\mathcal{D}_3(\ell)) = \sum_{d\mid\ell} d^2\,\sigma_1(\ell/d) = \ell^2 \sum_{d\mid\ell} \sigma_1(d)/d^2\,.$$

We can associate another zeta-function to the pair $(\mathcal{U}_n, \mathrm{Inv}(n;\mathbb{Z}))$.

(5.4) Lemma. *Given* $s\in\mathbb{C}$ *with* $\mathrm{Re}(s) > n$ *one has*

$$\zeta_n^*(s) = \sum_{A:\mathcal{U}_n\backslash\mathrm{Inv}(n;\mathbb{Z})} |\det A|^{-s} = \prod_{\ell=0}^{n-1}\zeta(s-\ell)\,.$$

Proof. According to (5.1) we calculate

$$\zeta_n^*(s) = \sum_{b_1,\ldots,b_n\in\mathbb{N}} b_1^{\,n-1-s}\cdot\ldots\cdot b_n^{\,-s} = \prod_{\ell=0}^{n-1}\zeta(s-\ell)\,. \qquad \square$$

(5.5) Exercises. a) Define the arithmetical function

$$d_n(\ell) := \#\,(\mathcal{U}_n\backslash\mathcal{D}_n(\ell)) \quad \text{for } \ell\in\mathbb{N}\,.$$

Show that d_n is multiplicative, i.e., $d_n(\ell m) = d_n(\ell)\,d_n(m)$ for relatively prime $\ell, m\in\mathbb{N}$. Verify (3) and show for every prime p that

$$d_n(p) = \frac{p^n-1}{p-1}\,,\quad d_n(p^2) = \frac{(p^n-1)\,(p^{n+1}-1)}{(p-1)\,(p^2-1)}\,.$$

b) Given $q\in\mathbb{N}$ show that the set

$$\left\{A\in\mathrm{Inv}(n;\mathbb{Z})\mid qA^{-1}\in\mathrm{Mat}(n;\mathbb{Z})\right\}$$

decomposes into full double cosets relative to $\mathcal{U}_n$ and consists of finitely many right cosets relative to $\mathcal{U}_n$.

c) Given $A\in\mathrm{Inv}(n;\mathbb{Z})$ show that $\mathcal{U}_n\cap A^{-1}\mathcal{U}_nA$ is a subgroup of $\mathcal{U}_n$ with finite index. Show that there exists $q\in\mathbb{N}$ such that $\mathcal{U}_n(q)$ (cf. (1.2)) is contained in this subgroup.

d) Recall the definition of a maximal parabolic subgroup from section III.5 . Then set $\mathcal{P}(m,n;\mathbb{Z}) := \mathcal{P}(m,n;\mathbb{Q})\cap\mathcal{U}_{m+n}$. Show that $(\mathcal{P}(m,n;\mathbb{Z}),\mathcal{P}(m,n;\mathbb{Q}))$ is a Hecke pair.

e) Derive the analogue of (5.1) for an arbitrary p.i.d. R. If K denotes the quotient field of R, show that $(GL(n;R),GL(n;K))$ is a Hecke pair, whenever R is norm-finite.

6. The Hecke Algebra Associated with $(\mathcal{U}_n, \mathrm{Inv}(n;\mathbb{Z}))$

We investigate the Hecke algebra

$$\mathcal{H}_n := \mathcal{H}(\mathcal{U}_n, \mathrm{Inv}(n;\mathbb{Z})) \quad \text{over} \quad R := \mathbb{Z}\ .$$

(6.1) Lemma. a) $\mathcal{H}_n$ *is a commutative algebra.*
b) *Given* $A \in \mathrm{Inv}(n;\mathbb{Z})$ *and* $0 \neq \lambda \in \mathbb{Z}$ *one has*

$$(\mathcal{U}_n A \mathcal{U}_n) \cdot (\mathcal{U}_n(\lambda I)\mathcal{U}_n) = \mathcal{U}_n(\lambda A)\mathcal{U}_n\ .$$

c) *Given* $A, B \in \mathrm{Inv}(n;\mathbb{Z})$ *such that* $\det A$ *and* $\det B$ *are relatively prime, one has*

$$(\mathcal{U}_n A \mathcal{U}_n) \cdot (\mathcal{U}_n B \mathcal{U}_n) = \mathcal{U}_n(AB)\mathcal{U}_n\ .$$

Proof. a) Transposing is an anti-automorphism of $GL(n;\mathbb{Q})$ satisfying

$$\mathcal{U}_n A' \mathcal{U}_n = \mathcal{U}_n A \mathcal{U}_n \quad \text{for all} \quad A \in \mathrm{Inv}(n;\mathbb{Z})$$

due to (3.3) . Now apply I(7.2) .
b) Use I(4.6) .
c) (4.5) yields $\mathcal{U}_n A \mathcal{U}_n B \mathcal{U}_n = \mathcal{U}_n AB \mathcal{U}_n$, hence

$$(\mathcal{U}_n A \mathcal{U}_n) \cdot (\mathcal{U}_n B \mathcal{U}_n) = t\, \mathcal{U}_n(AB)\mathcal{U}_n$$

for some $t \in \mathbb{N}$. In order to prove $t = 1$, choose

$$\iota(\mathcal{U}_n A \mathcal{U}_n) = \sum \mathcal{U}_n A_i\ , \quad \iota(\mathcal{U}_n B \mathcal{U}_n) = \sum \mathcal{U}_n B_j\ .$$

It suffices to demonstrate that the right cosets $\mathcal{U}_n A_i B_j$ are mutually distinct. Now let $U \in \mathcal{U}_n$ such that

$$UA_iB_j = A_kB_\ell\ , \quad \text{hence} \quad A_k^{-1}UA_i = B_\ell B_j^{-1}\ .$$

The denominators of the entries of the left hand side are divisors of $\det A$ and on the right hand side of $\det B$. Since the determinants are relatively prime, both sides are integral and therefore unimodular. Thus $\mathcal{U}_n B_\ell = \mathcal{U}_n B_j$ and $\mathcal{U}_n A_k = \mathcal{U}_n A_i$ lead to $j = \ell$ as well as $i = k$. □

It also follows from I(7.2) that the numbers of left and right cosets in each double coset relative to $\mathcal{U}_n$ coincide.

We derive a first consequence of the multiplicativity in (6.1c) . Given $\ell \in \mathbb{N}$ set

$$T_n(\ell) := \sum_{A:\mathcal{U}_n \backslash \mathcal{D}_n(\ell)/\mathcal{U}_n} \mathcal{U}_n A \mathcal{U}_n\ ,$$

cf. (5.3) . Then the multiplicativity of these elements $T_n(\ell)$ is stated in

(6.2) Corollary. *Given relatively prime* $k,\ell \in \mathbb{N}$ *, one has*

$$T_n(k) \cdot T_n(\ell) = T_n(k\ell) .$$

Proof. The elementary divisor matrices with determinant $k\ell$ coincide exactly with all the products of the elementary divisor matrices of determinant k and ℓ. Then apply (6.1) . □

(6.3) Remark. We sketch an alternative proof of (6.2) , (6.1c) , as well as (4.5) , which is due to A.N. Andrianov [4] , 3.2.5 . Let $k,\ell \in \mathbb{N}$ be relatively prime. Then we obtain

$$\text{(1)} \qquad \operatorname{ind} T_n(k\ell) = \operatorname{ind} T_n(k) \operatorname{ind} T_n(\ell) ,$$

because the Dirichlet series

$$\zeta_n^*(s) = \sum_{m=1}^{\infty} \operatorname{ind} T_n(m)\, m^{-s}$$

has an Euler product expansion due to (5.4) . Then the multiplicativity of the coefficients of this Dirichlet series is equivalent to (1) .

Given $A \in \mathcal{D}_n(k)$ and $B \in \mathcal{D}_n(\ell)$ we write

$$\text{(2)} \qquad \mathcal{U}_n A \mathcal{U}_n \cdot \mathcal{U}_n B \mathcal{U}_n = \mathcal{U}_n AB \mathcal{U}_n + \sum_{C:\mathcal{U}_n \backslash \mathcal{D}_n(k\ell)/\mathcal{U}_n} \alpha(A,B;C)\, \mathcal{U}_n C \mathcal{U}_n$$

with non-negative $\alpha(A,B;C) \in \mathbb{Z}$. The products of elementary divisor matrices of determinants k and ℓ are exactly the elementary divisor matrices of determinant $k\ell$. Thus

$$T_n(k) \cdot T_n(\ell) = T_n(k\ell) + \sum \alpha(A,B;C) \mathcal{U}_n C \mathcal{U}_n ,$$

holds, where A resp. B resp. C runs through the elementary divisor matrices of determinant k resp. ℓ resp. $k\ell$. Now I(4.5) and (1) imply $\alpha(A,B;C) = 0$ for all the coefficients appearing in (2) . This proves the assertions mentioned above.

Again we decompose the Hecke algebra into its primary components. Given a prime p consider the submonoid

$$\operatorname{Inv}(n;\mathbb{Z})_p := \{A \in \operatorname{Mat}(n;\mathbb{Z}) \mid |\det A| = p^\ell , \ \ell \in \mathbb{N}_0 \}$$

of $\operatorname{Inv}(n;\mathbb{Z})$ and the subalgebra

$$\mathcal{H}_{n,p} := \mathcal{H}(\mathcal{U}_n, \operatorname{Inv}(n;\mathbb{Z})_p) .$$

$\mathcal{H}_{n,p}$ is called the p-*primary component of* $\mathcal{H}_n$.

(6.4) Theorem. *The Hecke algebra* $\mathcal{H}_n$ *coincides with the tensor product of its primary components, i.e.,*

$$\mathcal{H}_n = \bigotimes_{p \text{ prime}} \mathcal{H}_{n,p} .$$

Proof. A basis of $\mathcal{H}_n$ resp. $\mathcal{H}_{n,p}$ is given by the double cosets. Each double coset possesses a unique representative in elementary divisor normal form. Each such matrix can be written as a finite product of such matrices, whose entries are powers of a fixed prime. Now (6.1c) completes the proof. □

In order to investigate $\mathcal{H}_n$ it therefore suffices to deal with the primary components $\mathcal{H}_{n,p}$.

The case $n = 1$ of (6.4) coincides with the Fundamental Theorem of Arithmetic. Each positive integer can be expressed as a finite product of prime powers, where the representation is unique up to the order.

(6.5) Exercises. a) Given $A \in \mathrm{Inv}(n;\mathbb{Z})$ define

$$\nu(A) := [\mathcal{U}_n : \mathcal{U}_n \cap A^{-1}\mathcal{U}_n A]$$

(cf. (5.5)). Given $A,B \in \mathrm{Inv}(n;\mathbb{Z})$ and $U,V \in \mathcal{U}_n$ show that

$$\nu(UAV) = \nu(A) \quad \text{and} \quad \nu(AB) = \nu(A)\,\nu(B) \quad , \text{ if } \quad \text{g.c.d.}(\det A, \det B) = 1 .$$

b) Given $q \in \mathbb{N}$ consider the set

$$M_n(q) := \{ A \in \mathrm{Inv}(n;\mathbb{Z}) \mid qA^{-1} \in \mathrm{Mat}(n,\mathbb{Z}) \} = \{ A \in \mathrm{Inv}(n;\mathbb{Z}) \mid \varepsilon_n(A) \mid q \}$$

(cf. (3.8)) and the element

$$S_n(q) := \sum_{A : \mathcal{U}_n \backslash M_n(q) / \mathcal{U}_n} \mathcal{U}_n A \mathcal{U}_n$$

in $\mathcal{H}_n$. Given relatively prime $k, \ell \in \mathbb{N}$ show that

$$S_n(k) \cdot S_n(\ell) = S_n(k\ell) .$$

c) Given $m,n \in \mathbb{N}$ show that the Hecke algebra $\mathcal{H}(\mathcal{P}(m,n;\mathbb{Z}),\mathcal{P}(m,n;\mathbb{Q}))$ fails to be commutative and contains zero-divisors (cf. (5.5) and IV(5.1)).

d) Derive the analogues for $\mathcal{H}(GL(n;R),\mathrm{Inv}(n;R))$, where R is a norm-finite p.i.d. .

7. Particular Double Cosets in the Primary Component

Fix a prime number p. In view of the Elementary Divisor Theorem each double coset $\mathcal{U}_n A \mathcal{U}_n$, $A \in \mathrm{Inv}(n;\mathbb{Z})_p$, possesses a unique representative of the form

(1) $$E(A) = \mathrm{diag}\left(p^{k_1},\dots,p^{k_n}\right) , \quad 0 \le k_1 \le \dots \le k_n .$$

Given $0 \le j \le n$ consider the double coset

$$T_{n,j} := T_{n,j}(p) := \mathcal{U}_n \,\mathrm{diag}\,(\underbrace{1,\dots,1}_{n-j},\underbrace{p,\dots,p}_{j})\,\mathcal{U}_n .$$

Given an integral matrix A define the *p-rank* of A by

$$r_p(A) := \text{rank } A \text{ over } \mathbb{Z}/p\mathbb{Z}.$$

Since unimodular matrices can be viewed as elements of $GL(n;\mathbb{Z}/p\mathbb{Z})$, we get

(2) $$r_p(A) = r_p(UAV) \text{ for unimodular } U,V.$$

Now (2) and the particular representative (1) lead to

(7.1) Lemma. *Given* $A \in \mathrm{Mat}(n;\mathbb{Z})$ *the following conditions are equivalent:*

(i) $A \in T_{n,j}(p)$.

(ii) $r_p(A) = n-j$ *and* $|\det A| = p^j$.

Given $1 \le j \le n$ the *j-th elementary symmetric polynomial* in the indeterminates $X_1,\dots,X_n$ is defined by

$$E_j(X_1,\dots,X_n) := \sum_{1\le\nu_1<\dots<\nu_j\le n} X_{\nu_1}\cdot\ldots\cdot X_{\nu_j}, \quad E_0(X_1,\dots,X_n) := 1.$$

Now we give a description of a set of representatives of the right cosets in $T_{n,j}(p)$.

(7.2) Proposition. *Given* $0 \le j \le n$ *a set of representatives of the right cosets relative to* $\mathcal{U}_n$ *in* $T_{n,j}(p)$ *is given by the lower triangular matrices*

(3) $$A = \begin{pmatrix} p^{k_1} & & 0 \\ & \ddots & \\ a_{\mu\nu} & & p^{k_n} \end{pmatrix},$$

which satisfy

(i) $k_\nu \in \{0,1\}$ *for* $\nu = 1,\dots,n$,

(ii) $k_1+\dots+k_n = j$,

(ii) $0 \le a_{\mu\nu} < p^{k_\nu(1-k_\mu)}$ *for* $1 \le \nu < \mu \le n$.

In particular one has

$$\text{ind } T_{n,j}(p) = p^{-j(j+1)/2}\, E_j(p,p^2,\dots,p^n).$$

Proof. If A satisfies (i), (ii), (iii), one has $\det A = p^j$ and $r_p(A) = n-j$, hence (7.1) yields $A \in T_{n,j}(p)$. Since the matrices A have the form (5.1), the right cosets are mutually distinct.

Conversely choose a set of representatives of $\mathcal{U}_n\backslash T_{n,j}(p)$ in the form (5.1). These matrices have the form (3) and satisfy $0 \le a_{\mu\nu} < p^{k_\nu}$. Then (ii) follows from the condition on the determinant and (i) holds, because pA^{-1} turns out to be integral again for all matrices $A \in T_{n,j}(p)$.

In order to demonstrate (iii) assume that there exists a pair (μ,ν) such that

$$a_{\mu\nu} \neq 0\,,\quad k_\mu = k_\nu = 1\,,\quad 1 \le \nu < \mu \le n\,.$$

Then there exist at least $n-j+1$ over $\mathbb{Z}/p\mathbb{Z}$ linearly independent rows in A, namely those $n-j$ with $k_\ell = 0$ and the μ-th row. This contradicts (7.1).

If $k_1,\dots,k_n$ are fixed, there are

$$\prod_{1\le\nu<\mu\le n} p^{k_\nu(1-k_\mu)}$$

possibilities for A. Next $k_\nu^2 = k_\nu$ implies

$$\frac{1}{2}\,j(j+1) = \frac{1}{2}\Big(\sum_{\nu=1}^{n} k_\nu\Big)\Big(1+\sum_{\mu=1}^{n} k_\mu\Big) = \sum_{\nu=1}^{n} k_\nu + \sum_{1\le\nu<\mu\le n} k_\mu k_\nu\,.$$

This leads to

$$p^{-j(j+1)/2}\prod_{\nu=1}^{n} p^{(n+1-\nu)k_\nu} = \prod_{1\le\nu<\mu\le n} p^{k_\nu(1-k_\mu)}\,.$$

Finally we observe that

$$\begin{aligned}\operatorname{ind} T_{n,j}(p) &= p^{-j(j+1)/2}\sum_{\substack{(k_1,\dots,k_n)'\in\{0,1\}^n\\ k_1+\dots+k_n=j}} p^{nk_1}\,p^{(n-1)k_2}\cdot\ldots\cdot p^{k_n}\\ &= p^{-j(j+1)/2}\,E_j\big(p,p^2,\dots,p^n\big)\,.\end{aligned}$$

□

In particular one has

$$T_{n,0}(p) = \mathcal{U}_n\,,\qquad \operatorname{ind} T_{n,0}(p) = 1\,,$$

$$T_{n,n}(p) = \mathcal{U}_n(pI)\mathcal{U}_n\,,\qquad \operatorname{ind} T_{n,n}(p) = 1\,.$$

In the case $n = 2$ we observe

$$T_{2,1}(p) = \mathcal{U}_2\begin{pmatrix}1&0\\0&p\end{pmatrix}\mathcal{U}_2\,,\qquad \operatorname{ind} T_{2,1}(p) = p+1 = \sigma_1(p)\,.$$

In the case $n = 3$ we have

$$\operatorname{ind} T_{3,1}(p) = \operatorname{ind} T_{3,2}(p) = \frac{p^3-1}{p-1}\,.$$

(7.3) Remark. a) Given $0 \le j \le n$ the mapping

$$\begin{aligned}\mathcal{U}_n\backslash T_{n,j}(p) &\to \mathcal{U}_n\backslash T_{n,n-j}(p)\,,\\ \mathcal{U}_n A &\mapsto \mathcal{U}_n pA'^{-1}\,,\end{aligned}$$

is a bijection. This follows from (1) and (7.1). In particular one has

$$\operatorname{ind} T_{n,j}(p) = \operatorname{ind} T_{n,n-j}(p)\,.$$

b) One easily checks the following identity for $0 \le j \le n$

$$(4) \qquad p^{-j(j+1)/2}\, E_j(p,p^2,\dots,p^n) = \frac{\prod_{\ell=1}^{n}(p^\ell-1)}{\prod_{\ell=1}^{j}(p^\ell-1)\prod_{\ell=1}^{n-j}(p^\ell-1)} = \operatorname{ind} T_{n,j}(p) \ ,$$

where the vacuous product is defined to be 1 . Here (4) equals the number of j-dimensional subspaces of $(\mathbb{Z}/p\mathbb{Z})^n$ (cf. [58] , Proposition 3.18) .

(7.4) Exercises. a) Prove Remark (7.3) .

b) Describe the analogous set of representatives of $\mathcal{U}_n\backslash T_{n,j}(p)$, which consists of upper triangular matrices (cf. (7.2)) .

c) Show that

$$S_n(p) = T_{n,0}(p) + T_{n,1}(p) + \dots + T_{n,n}(p)$$

holds (cf. (6.5)) .

d) Derive the analogous results for the π-primary component of the Hecke algebra $\mathcal{H}(GL(n;R),Inv(n;R))$, where R is a norm-finite p.i.d. and π a prime element in R .

8. The Primary Component

In this section we determine the structure of the primary components. Fix a prime p . We observe that $\mathcal{H}_{n,p}$ is a $\mathbb{Z}$-graded algebra. One has

$$\mathcal{H}_{n,p} = \bigoplus_{k=0}^{\infty} \mathcal{H}_{n,p}^{(k)} \ ,$$

where

$$\mathcal{H}_{n,p}^{(k)} := \left\{ \sum_{A:\mathcal{U}_n\backslash\mathcal{D}(p^k)/\mathcal{U}_n} t(\mathcal{U}_nA\mathcal{U}_n)\,\mathcal{U}_nA\mathcal{U}_n \mid t(\mathcal{U}_nA\mathcal{U}_n) \in \mathbb{Z} \right\}.$$

This means

$$(1) \qquad \mathcal{H}_{n,p}^{(0)} = \mathbb{Z}\,\mathcal{U}_n \ ,$$

$$(2) \qquad \mathcal{H}_{n,p}^{(1)} = \mathbb{Z}\,T_{n,1}(p) \ ,$$

$$(3) \qquad \mathcal{H}_{n,p}^{(k)} \cdot \mathcal{H}_{n,p}^{(\ell)} \subset \mathcal{H}_{n,p}^{(k+\ell)} \ .$$

A result on the generators of the Hecke algebra is derived in

(8.1) Proposition. *The* $\mathbb{Z}$*-algebra* $\mathcal{H}_{n,p}$ *is generated by*

$$T_{n,0}(p),\dots,T_{n,n}(p) \ .$$

Proof. Let $\mathcal{A}$ denote the subalgebra of $\mathcal{H}_{n,p}$ generated by $T_{n,0},\dots,T_{n,n}$. We show by induction on k that

(4) $$\mathcal{H}_{n,p}^{(k)} \subset \mathcal{A}$$

holds. Here $k = 0,1$ are clear in view of (1), (2).

Given $k \geq 2$ we show successively that the submodule

$$\mathcal{H}_{n,p}^{(k,r)} := \Big\{ \sum_{\substack{A:\mathcal{U}_n\backslash\mathcal{D}_n(p^k)/\mathcal{U}_n \\ r_p(A)=r}} t(\mathcal{U}_n A \mathcal{U}_n)\, \mathcal{U}_n A \mathcal{U}_n \mid t(\mathcal{U}_n A \mathcal{U}_n) \in \mathbb{Z} \Big\}$$

is contained in $\mathcal{A}$. Given a double coset $\mathcal{U}_n A \mathcal{U}_n \in \mathcal{H}_{n,p}^{(k,r)}$ choose the representative A in the form

(5) $$A = \operatorname{diag}(p^{k_1},\dots,p^{k_n}), \quad 0 = k_1 = \dots = k_r < k_{r+1} \leq \dots \leq k_n, \quad k_1 + \dots + k_n = k.$$

If $r = 0$ one has $k_1 \geq 1$. Then

$$T := \mathcal{U}_n(\tfrac{1}{p}A)\mathcal{U}_n \in \mathcal{A}$$

holds by induction hypothesis. In view of $T_{n,n} = \mathcal{U}_n(pI)\mathcal{U}_n$ now (6.1) leads to

$$\mathcal{U}_n A \mathcal{U}_n = T \cdot T_{n,n} \in \mathcal{A}.$$

Since $r = n$ corresponds to $k = 0$, we may assume $0 < r < n$. Now consider

$$T := \mathcal{U}_n \operatorname{diag}(1,\dots,1,p^{k_{r+1}-1},\dots,p^{k_n-1})\mathcal{U}_n \in \mathcal{H}_{n,p}^{(k+r-n)} \subset \mathcal{A}.$$

We compute

$$T \cdot T_{n,n-r} = \sum_{B:\mathcal{U}_n\backslash\mathcal{D}_n(p^k)/\mathcal{U}_n} t(\mathcal{U}_n B \mathcal{U}_n)\, \mathcal{U}_n B \mathcal{U}_n \in \mathcal{H}_{n,p}^{(k)}.$$

At first $t(\mathcal{U}_n B \mathcal{U}_n) \neq 0$ implies $r_p(B) \leq r$ in view of (7.1). Consider $\mathcal{U}_n B \mathcal{U}_n \in \mathcal{H}_{n,p}^{(k,r)}$ with a representative

$$B = \operatorname{diag}(1,\dots,1,p^{\ell_{r+1}},\dots,p^{\ell_n}), \quad 1 \leq \ell_{r+1} \leq \dots \leq \ell_n, \quad \ell_{r+1} + \dots + \ell_n = k.$$

Choose the representatives of the right cosets $\mathcal{U}_n\backslash T_{n,n-r}$ in the form (5.1). Then the quantity $t(\mathcal{U}_n B \mathcal{U}_n)$ equals the number of C's such that

$$BC^{-1} \in T.$$

But the choice of B immediately yields

$$C = \begin{pmatrix} I^{(r)} & 0 \\ 0 & pI^{(n-r)} \end{pmatrix}.$$

Then $BC^{-1} \in T$ holds if and only if $BC^{-1} = \operatorname{diag}(1,\dots,1,p^{k_{r+1}-1},\dots,p^{k_n-1})$, because BC^{-1} has elementary divisor normal form. Then $B = A$ follows. Thus we have

$$T \cdot T_{n,r} = T_1 + \mathcal{U}_n A \mathcal{U}_n,$$

where

$$T_1 \in \bigoplus_{\ell=0}^{r-1} \mathcal{H}_{n,p}^{(k,\ell)} .$$

The induction hypothesis yields

$$\mathcal{A} = \mathcal{H}_{n,p} . \qquad \square$$

Note that this proof does not use the explicit description of the representatives of the right cosets in $T_{n,j}(p)$ given in (7.2), but only (5.1) and (7.1).

The structure of the primary component is derived in

(8.2) Theorem. *$\mathcal{H}_{n,p}$ equals the polynomial ring over $\mathbb{Z}$ in n elements*

$$T_{n,1}(p),\ldots,T_{n,n}(p) ,$$

which are algebraically independent. In particular $\mathcal{H}_n$ does not contain any zero-divisors.

Proof. $$\mathcal{H}_{n,p} = \mathbb{Z}[T_{n,1}(p),\ldots,T_{n,n}(p)]$$

follows from (8.1). In view of the elementary divisor normal form (1) in section 7, we calculate

$$\dim \mathcal{H}_{n,p}^{(k)} = \#\{(k_1,\ldots,k_n)' \in \mathbb{Z}^n \mid 0 \le k_1 \le \ldots \le k_n,\ k_1+\ldots+k_n = k\} = p_n(k) ,$$

where $p_n(k)$ denotes the number of partitions of k into at most n summands. Due to (3) a monomial

$$T_{n,1}^{\ell_1} \cdot\ldots\cdot T_{n,n}^{\ell_n}$$

belongs to $\mathcal{H}_{n,p}^{(k)}$ if and only if

$$\ell_1 + 2\ell_2 + \ldots + n\ell_n = k .$$

According to [5], Theorem 14.1, the number of these monomials also equals $p_n(k)$. Since the monomials span $\mathcal{H}_{n,p}^{(k)}$, they are linearly independent. Hence $T_{n,1},\ldots,T_{n,n}$ are algebraically independent. $\square$

It is easy to derive the structure of the Hecke algebra $\bar{\mathcal{H}} = \mathcal{H}(\mathcal{U}_n, GL(n;\mathbb{Q}))$. One obtains

$$\bar{\mathcal{H}}_n = \bigotimes_{p \text{ prime}} \bar{\mathcal{H}}_{n,p} ,$$

where

$$\bar{\mathcal{H}}_{n,p} = \mathbb{Z}[T_{n,1}(p), T_{n,2}(p),\ldots,T_{n,n}(p), T_{n,n}(p)^{-1}] .$$

G. Shimura [58] proves (8.2) by induction on n. Therefore he needs a homomorphism, which will be described in the sequel. Given $n > 1$ then each right coset $\mathcal{U}_n A$,

$A \in \mathrm{Inv}(n;\mathbb{Z})$, contains a representative of the form

(6) $$A = \begin{pmatrix} a_1 & 0 \\ a & A_1 \end{pmatrix}, \quad a_1 \in \mathbb{N}, \quad a \in \mathbb{Z}^{n-1}, \quad A_1 \in \mathrm{Inv}(n-1;\mathbb{Z}).$$

Now we define

$$\Phi(\mathcal{U}_n A) = \begin{Bmatrix} \mathcal{U}_{n-1} A_1 & \text{if } a_1 = 1 \\ 0 & \text{if } a_1 > 1 \end{Bmatrix} \in \mathcal{R}_{n-1} = \mathcal{R}(\mathcal{U}_{n-1}, \mathrm{Inv}(n-1,\mathbb{Z})).$$

Clearly the definition does not depend on the choice of the representative A as long as A has the form (6) . If $a_1 = 1$, one may choose $a = 0$. Φ is linearly extended to $\mathcal{R}_n$. Multiplication by matrices $\begin{pmatrix} 1 & 0 \\ 0 & U_1 \end{pmatrix}$, $U_1 \in \mathcal{U}_{n-1}$, from the right hand side yields

$$\Phi\left(\mathcal{R}_n^{\mathcal{U}_n}\right) \subset \mathcal{R}_{n-1}^{\mathcal{U}_{n-1}}, \text{ hence } \Phi(\mathcal{H}_n) \subset \mathcal{H}_{n-1}$$

due to I(3.9) . The definition of the action immediately leads to

$$\Phi(T * U) = \Phi(T) * \Phi(U) \quad \text{for } T \in \mathcal{H}_n, \; U \in \mathcal{R}_n.$$

Thus Φ turns out to be a homomorphism of the Hecke algebras.

(8.3) Lemma. *If* $n > 1$, *the homomorphism*

$$\Phi : \mathcal{H}_n \to \mathcal{H}_{n-1}, \quad T \mapsto \Phi(T),$$

is surjective. Given a prime p *and* $0 \le j < n$, *one has*

(7) $$\Phi(T_{n,j}(p)) = T_{n-1,j}(p).$$

Moreover the kernel of $\Phi|\mathcal{H}_{n,p}$ *is a principal ideal generated by* $T_{n,n}(p)$.

Proof. (7) follows from (7.2) . Hence Φ proves to be surjective due to (8.2) combined with (6.4) . Clearly $T_{n,n}(p) \in \text{kernel } \Phi$ holds. If we combine this with (7) and (8.2) , we get

$$\text{kernel } \Phi|\mathcal{H}_{n,p} = T_{n,n}(p) \cdot \mathcal{H}_{n,p}. \qquad \square$$

Note that this proof makes use of the explicit description of the representatives of the right coset in (7.2) .

Another homomorphism concerns the case, where $\mathcal{U}_n$ is replaced by $\mathcal{SU}_n = SL(n;\mathbb{Z})$. Let $\mathrm{Inv}^+(n;\mathbb{Z}) = \{A \in \mathrm{Mat}(n;\mathbb{Z}) \mid \det A > 0\}$. Then we obviously obtain

(8.4) Remark. The Hecke algebras $\mathcal{H}(\mathcal{SU}_n, \mathrm{Inv}^+(n;\mathbb{Z}))$ and $\mathcal{H}(\mathcal{U}_n, \mathrm{Inv}(n;\mathbb{Z}))$ are canonically isomorphic.

(8.5) Exercises. a) Show that the Hecke algebra $\mathcal{H}(\mathcal{SU}_3, SL(3;\mathbb{Q}))$ over $R = \mathbb{Z}$ consists of the polynomials over $\mathbb{Z}$ in the algebraically independent elements

$$\mathcal{SU}_3\operatorname{diag}(\tfrac{1}{p},1,p)\mathcal{SU}_3\ ,\quad \mathcal{SU}_3\operatorname{diag}(\tfrac{1}{p},\tfrac{1}{p},p^2)\mathcal{SU}_3\ ,$$

where p runs through all primes.

b) Express $\mathcal{U}_3\operatorname{diag}(p,p^2,p^3)\mathcal{U}_3$ as a polynomial in $T_{3,1}(p)$, $T_{3,2}(p)$, and $T_{3,3}(p)$.

c) Derive the analogous results for the π-primary component of the Hecke algebra $\mathcal{H}(GL(n;R),Inv(n;R))$, where R is a norm-finite p.i.d. .

9. Formal Power Series

In this section we will prove the analogue of IV(4.3) for arbitrary degree n , where the result is due to T. Tamagawa [62] . We will show that a particular formal power series over the Hecke algebra $\mathcal{H}_n$ given by determinantal conditions is in fact a rational function. Therefore we compute a particular product of the generating double cosets with the formal sum of all double cosets of fixed determinant.

Given $\ell \in \mathbb{N}$ recall the definition

$$T_n(\ell) := \sum_{A:\mathcal{U}_n\backslash\mathcal{D}_n(\ell)/\mathcal{U}_n} \mathcal{U}_n A \mathcal{U}_n\ ,$$

cf. (5.3) . Let p be a prime, $\ell \geq 0$, and $0 \leq r \leq n$. Then set

$$T_n^{(r)}(p^\ell) = \sum_{\substack{A:\mathcal{U}_n\backslash\mathcal{D}_n(p^\ell)/\mathcal{U}_n \\ r_p(A)=r}} \mathcal{U}_n A \mathcal{U}_n\ .$$

Given $n > 1$ one can apply the Φ-operator (cf. (8.3)) and obtains

$$\Phi(T_n(\ell)) = T_{n-1}(\ell)$$

or more precisely for $r > 0$

$$\Phi\left(T_n^{(r)}(p^\ell)\right) = T_{n-1}^{(r-1)}(p^\ell)\ .$$

Note that

(1)
$$\begin{aligned}
T_n^{(r)}(p^\ell) &= 0\ ,\ \text{if}\ r+\ell < n\ ,\\
T_n(p^\ell) &= \sum_{r=0}^{n} T_n^{(r)}(p^\ell)\ ,\\
T_n(1) &= \mathcal{U}_n\ ,\quad T_n(p) = T_{n,1}(p)\ ,\\
T_{n,j}(p) &= T_n^{(n-j)}(p^j)\ ,\quad 0 \leq j \leq n\ .
\end{aligned}$$

The products of the generating double cosets with $T_n(p^\ell)$ are computed in

(9.1) Proposition. *Given a prime* p , $1 \leq j \leq n$, *and* $\ell \in \mathbb{N}$, *one has*

$$T_n(p^\ell)\cdot T_{n,j}(p) = \sum_{r=0}^{n-j} \left(\operatorname{ind} T_{n-r,j}(p)\right) T_n^{(r)}(p^{\ell+j})\ .$$

Proof. Set

$$T_n(p^\ell)\cdot T_{n,j}(p) = \sum_{B:\mathcal{U}_n\backslash\mathcal{D}_n(p^\ell)/\mathcal{U}_n} t(\mathcal{U}_nB\mathcal{U}_n)\,\mathcal{U}_nB\mathcal{U}_n .$$

Clearly $t(\mathcal{U}_nB\mathcal{U}_n)\neq 0$ implies $r_p(B)\le n-j$ according to (7.1) . Let $r = r_p(B)\le n-j$ and choose the representative B in the form

$$B = \operatorname{diag}\left(1,\ldots 1,p^{\nu_{r+1}},\ldots,p^{\nu_n}\right),\quad 1\le \nu_{r+1}\le\ldots\le\nu_n ,\quad \nu_{r+1}+\ldots+\nu_n = \ell+j .$$

Consider the representatives A of $\mathcal{U}_n\backslash T_{n,j}(p)$ as in (7.2) . Then $t(\mathcal{U}_nB\mathcal{U}_n)$ equals the number of A 's , for which BA^{-1} is integral, since they then obviously belong to $\mathcal{D}_n(p^\ell)$. Now BA^{-1} proves to be integral if and only if A has the form

$$\begin{pmatrix} I & 0\\ 0 & A_1\end{pmatrix},\quad I = I^{(r)} ,$$

where A_1 runs through a set of representatives of $\mathcal{U}_{n-r}\backslash T_{n-r,j}(p)$ by virtue of (7.2) , since pA_1^{-1} is always integral. Hence we get

$$t(\mathcal{U}_nB\mathcal{U}_n) = \operatorname{ind} T_{n-r,j}(p) . \qquad \square$$

Furthermore we need a result on result on certain sums of elementary symmetric polynomials.

(9.2) Lemma. *If* X *is an indeterminate, one has for all* $n\ge 0$

$$\sum_{\ell=0}^{n} (-1)^\ell E_\ell\left(1,X,\ldots,X^{n-1}\right) = \begin{cases} 1 & \text{if } n = 0 ,\\ 0 & \text{if } n > 0 .\end{cases}$$

Proof. The cases $n = 0,1$ are obvious. Given $n\ge 2$ calculate

$$\sum_{\ell=0}^{n} (-1)^\ell E_\ell\left(1,X,\ldots,X^{n-1}\right) = \left(1-X^{n-1}\right)\sum_{\ell=0}^{n-1} (-1)^\ell E_\ell\left(1,X,\ldots,X^{n-2}\right) = 0 . \qquad \square$$

As a generalization of IV(4.3) we obtain

(9.3) Theorem. *Rationality Theorem*

Let X *be an indeterminate and* p *a prime. Then*

$$\sum_{\ell=0}^{\infty} T_n(p^\ell)\,X^\ell = \left(\sum_{j=0}^{n} (-1)^j p^{j(j-1)/2}\,T_{n,j}(p)\,X^j\right)^{-1}$$

holds formally in X .

Proof. In view of (7.2) we have

$$\operatorname{ind} T_{n,j}(p) = p^{-j(j-1)/2}\,E_j\left(1,p,\ldots,p^{n-1}\right) .$$

Now we use (9.1) and (1) in the following calculation

$$\left(\sum_{j=0}^{n} (-1)^j p^{j(j-1)/2} T_{n,j}(p)\, X^j\right) \cdot \left(\sum_{\ell=0}^{\infty} T_n(p^\ell)\, X^\ell\right)$$

$$= \sum_{j=0}^{n} (-1)^j p^{j(j-1)/2} \sum_{m=j}^{n} \operatorname{ind} T_{m,j}(p) \sum_{\ell=m-j}^{\infty} T_n^{(n-m)}(p^{\ell+j})\, X^{\ell+j}$$

$$= \sum_{m=0}^{n} \left(\sum_{j=0}^{m} (-1)^j E_j(1,p,\dots,p^{m-1})\right) \sum_{r=m}^{\infty} T_n^{(n-m)}(p^r)\, X^r .$$

Due to (9.2) all the summands vanish except for $m = 0$. Hence (1) leads to the result

$$\sum_{\ell=0}^{\infty} T_n^{(n)}(p^\ell)\, X^\ell = \mathcal{U}_n . \qquad \square$$

Finally a combination of (9.3) with (6.2) yields a formal Euler product. If we set $X = p^{-s}$, we get

(9.4) Corollary. *The identity*

$$\sum_{\ell=1}^{\infty} T_n(\ell)\, \ell^{-s} = \prod_{p \text{ prime}} \left(\sum_{j=0}^{n} (-1)^j T_{n,j}(p)\, p^{(j(j-1)/2)-js}\right)^{-1}$$

holds formally in s .

In the case $n = 3$ D. Bump [12] derives the following identity

(9.5) Corollary. *Let* X *be an indeterminate and* p *a prime. Then*

$$\sum_{\ell=0}^{\infty} T_3(p^\ell)\, X^\ell = \left(1 - T_3(p)\, X + \left(T_3(p)^2 - T_3(p^2)\right) X^2 - p^3\, T_{3,3}(p)\, X^3\right)^{-1}$$

holds formally in X .

Proof. (9.1) yields

$$T_3(p)^2 = (p+1)\, \mathcal{U}_3 \operatorname{diag}(1,p,p) \mathcal{U}_3 + \mathcal{U}_3 \operatorname{diag}(1,1,p^2) \mathcal{U}_3 = T_3(p^2) + p\, T_{3,2}(p) .$$

Now insert this into (9.3) . $\square$

(9.6) Exercises. a) Demonstrate (9.1) by use of the Φ-operator in (8.3) and an induction on n .

b) Given $n \ge 2$ demonstrate

$$T_n(p)^2 = T_n(p^2) + p\, T_{n,2}(p) .$$

c) Use (9.4) in order to demonstrate

$$\zeta_n^*(s) = \sum_{\ell=1}^{\infty} \operatorname{ind} T_n(\ell)\, \ell^{-s} = \prod_{\ell=0}^{n-1} \zeta(s-\ell)$$

for all $s \in \mathbb{C}$ with $\mathrm{Re}(s) > n$ (cf. (5.4)) .

d) Derive the analogous results for the π-primary component of the Hecke-algebra $\mathcal{H}(GL(n;R),Inv(n;R))$, where R is a norm-finite p.i.d. .

10. The Hecke Algebra over Q

In the applications it is much more convenient to deal with the elements $T_n(\ell)$ than with $T_{n,j}(p)$, since the right coset decompositions are simpler. This is posssible without loss of information if we replace the ground ring $\mathbb{Z}$ by Q or an arbitrary field of characteristic 0 .

Let $\tilde{\mathcal{H}}_n$ resp. $\tilde{\mathcal{H}}_{n,p}$ denote the corresponding Hecke algebras over Q , i.e.,

(1) $$\tilde{\mathcal{H}}_n = \mathcal{H}_n \otimes_{\mathbb{Z}} Q, \quad \tilde{\mathcal{H}}_{n,p} = \mathcal{H}_{n,p} \otimes_{\mathbb{Z}} Q .$$

Instead of (8.2) we also get

(10.1) Theorem. *Given a prime* p *one has*

$$\tilde{\mathcal{H}}_{n,p} = Q[T_n(p),T_n(p^2),\dots,T_n(p^n)] .$$

Proof. In view of (1) and (8.2) we conclude

$$\tilde{\mathcal{H}}_{n,p} = Q[T_{n,1}(p),\dots,T_{n,n}(p)] .$$

We apply (9.3) and compute the coefficient of X^m, $1 \le m \le n$, in

$$\left(\sum_{j=0}^{n} (-1)^j p^{j(j-1)/2} T_{n,j}(p) X^j\right)\cdot\left(\sum_{\ell=0}^{\infty} T_n(p^\ell) X^\ell\right) = \mathcal{U}_n .$$

Hence we get

$$0 = \sum_{j=0}^{m} (-1)^j p^{j(j-1)/2} T_{n,j}(p)\cdot T_n(p^{m-j})$$

or equivalently

$$(-1)^{m-1} p^{m(m-1)/2} T_{n,m}(p) = \sum_{j=0}^{m-1} (-1)^j p^{j(j-1)/2} T_{n,j}(p)\cdot T_n(p^{m-j}) .$$

In view of $T_{n,0}(p) = T_n(1) = \mathcal{U}_n$ we conclude successively that all the $T_{n,j}(p)$, $1 \le j \le n$, belong to the algebra

$$Q[T_n(p),\dots,T_n(p^n)] . \qquad \square$$

We show why we have to replace $\mathbb{Z}$ by Q .

(10.2) Remark. Over $R = \mathbb{Z}$ the polynomial ring $\mathbb{Z}[T_n(p),\dots,T_n(p^n)]$ becomes a proper subalgebra of $\mathcal{H}_n$, whenever $n > 1$. If $n = 2$, then IV(4.1) leads to

$$\mathbb{Z}[T_2(p),T_2(p^2)] = \mathbb{Z}[T_{2,1}(p),p\,T_{2,2}(p)] .$$

If we use (6.2) and

$$\tilde{\mathcal{H}}_n = \bigotimes_{p \text{ prime}} \tilde{\mathcal{H}}_{n,p} ,$$

(10.1) implies

(10.3) Corollary. *The Hecke algebra* $\tilde{\mathcal{H}}_n$ *is generated by* $T_n(\ell)$, $\ell \in \mathbb{N}$.

The so-called spherical mapping introduced by I. Satake [53] gives a description of $\tilde{\mathcal{H}}_{n,p}$ in terms of symmetric polynomials. Therefore let $X_1,\dots,X_n$ be independent indeterminates. Given $A \in \mathrm{Inv}(n;\mathbb{Z})_p$ we choose a representative of the right coset $\mathcal{U}_n A$ in the form (5.2) as a lower triangular matrix with $p^{k_1},\dots,p^{k_n}$ as diagonal entries. Then set

$$P(\mathcal{U}_n A) := \prod_{j=1}^{n} (p^{j-1-n} X_j)^{k_j} .$$

This map is extended to $\tilde{\mathcal{R}}_{n,p} = \mathcal{R}(\mathcal{U}_n, \mathrm{Inv}(n;\mathbb{Z})_p) \otimes_{\mathbb{Z}} \mathbb{Q}$ by linearity. Given $T \in \tilde{\mathcal{H}}_{n,p}$ set

$$P(T) := P(\iota(T)) .$$

Let $\mathbb{Q}[X_1,\dots,X_n]^{S_n}$ denote the algebra of symmetric polynomials. Then the result can be described as

(10.4) Theorem. *Given a prime* p *the map*

$$P : \tilde{\mathcal{H}}_{n,p} \rightarrow \mathbb{Q}[X_1,\dots,X_n]^{S_n}$$

is an isomorphism of the $\mathbb{Q}$*-algebras and satisfies*

$$(2) \qquad P(T_{n,j}(p)) = p^{-j(j+1)/2} E_j(X_1,\dots,X_n) , \quad 0 \le j \le n .$$

Proof. In view of the definition of the product

$$P : \tilde{\mathcal{H}}_{n,p} \rightarrow \mathbb{Q}[X_1,\dots,X_n]$$

is an injective homomorphism of the algebras. Since the elementary symmetric polynomials generate $\mathbb{Q}[X_1,\dots,X_n]^{S_n}$, it suffices to demonstrate (2) in view of (8.2) . Let $0 \le j \le n$. If $k_1,\dots,k_n$ are fixed, the number of representatives of $\mathcal{U}_n \backslash T_{n,j}(p)$, which are lower triangular matrices and have $p^{k_1},\dots,p^{k_n}$ as its diagonal entries equals

$$\prod_{1 \le \nu < \mu \le n} p^{k_\nu(1-k_\mu)} = p^{-j(j+1)/2} \prod_{\nu=1}^{n} p^{(n+1-\nu)k_\nu}$$

according to the proof of (7.2) . Then (7.2) immediately leads to (2) . □

The phenomenon that the symmetric polynomials appear in (10.4) can be explained by the fact that S_n is the Weyl group of the general linear group of degree n .

Using the homomorphism P the polynomial in the Rationality Theorem (9.3) decomposes into linear factors. A combination of (10.4) and (9.3) immediately leads to

(10.5) Corollary. *If* p *is a prime, one has*

$$\sum_{j=0}^{n} (-1)^j p^{j(j-1)/2} P(T_{n,j}(p)) X^j = \prod_{j=1}^{n} (1 - \tfrac{1}{p} X_j X) .$$

(10.6) Exercises. a) Let p be a prime. Show that $\mathbb{Z}[T_n(p),\dots,T_n(p^n)]$ is equal to the algebra of polynomials over $\mathbb{Z}$ in the elements

$$p^{j(j-1)/2} T_{n,j}(p) , \quad j = 1,\dots,n .$$

b) If p is a prime, determine $P\big(T_n(p^\ell)\big)$ for $\ell \geq 0$.

c) If p is a prime, determine $T \in \mathcal{H}_{n,p}$ such that $P(T) = X_1{}^2 + X_2{}^2 + \dots + X_n{}^2$.

d) Derive the analogous results for the Hecke algebra $\mathcal{H}(GL(n;R),Inv(n;R))$, where R is a norm-finite pid.

11. An Embedding

In this section we consider the analogue of IV.5 for the general case of n . Recall the definition of the Borel group $\mathcal{B}(n;\mathbb{Q})$ from section III.4 and set

$$\mathcal{B}_n = \mathcal{B}(n;\mathbb{Q}) \cap \mathcal{U}_n = \{ U \in \mathcal{U}_n \mid U \text{ upper triangular } \} ,$$

$$M_n = \mathcal{B}(n;\mathbb{Q}) \cap Mat(n;\mathbb{Z}) = \{ A \in Inv(n;\mathbb{Z}) \mid A \text{ upper triangular } \} .$$

Each right coset $\mathcal{B}_n A$, $A \in M_n$, clearly contains a unique representative of the form (5.2) . Due to (5.2) the conditions

$$\mathcal{U}_n \cap M_n = \mathcal{B}_n , \quad Inv(n;\mathbb{Z}) = \mathcal{U}_n M_n$$

are satisfied. We get an embedding

$$\nu : \mathcal{R}(\mathcal{U}_n, Inv(n;\mathbb{Z})) \to \mathcal{R}(\mathcal{B}_n, M_n) ,$$

$$\sum_{A:\mathcal{U}_n\backslash Inv(n;\mathbb{Z})} u(\mathcal{U}_n A)\mathcal{U}_n A \mapsto \sum_{A:\mathcal{B}_n\backslash M_n} u(\mathcal{U}_n A)\mathcal{B}_n A .$$

According to I(6.4) we obtain an injective homomorphism

$$\nu : \mathcal{H}(\mathcal{U}_n, Inv(n;\mathbb{Z})) \to \mathcal{H}(\mathcal{B}_n, M_n)$$

of the algebras. If $n > 1$ clearly $\mathcal{H}(\mathcal{B}_n, M_n)$ fails to be commutative and contains zero-divisors due to IV(5.1) . Given $1 \leq j \leq n$ consider the double coset

$$S_{n,j}(p) := \mathcal{B}_n diag(\underbrace{1,\dots,1}_{j-1},p,\underbrace{1,\dots,1}_{n-j})\mathcal{B}_n \in \mathcal{H}(\mathcal{B}_n, M_n) ,$$

A set of representatives of $\mathcal{B}_n \backslash S_{n,j}(p)$ is given by

$$(1)\qquad \begin{pmatrix} I & x & 0 \\ 0 & p & 0 \\ 0 & 0 & I \end{pmatrix}, \quad x = (x_1,\ldots,x_{j-1})', \quad 0 \le x_i < p, \quad 1 \le i \le j-1 .$$

(11.1) Lemma. *Given* $1 \le j \le n$ *and a prime* p, *one has*

$$p^{j(j-1)/2}\, \nu\big(T_{n,j}(p)\big) = \sum_{1 \le i_1 < \ldots < i_j \le n} S_{n,i_1}(p) \cdot S_{n,i_2}(p) \cdot \ldots \cdot S_{n,i_j}(p) .$$

Proof. A set of representatives of $\mathcal{U}_n \backslash T_{n,j}(p)$ in the form (5.2) is obtained from (7.2) by conjugating with

$$U = \begin{pmatrix} 0 & & 1 \\ & \cdot^{\cdot^{\cdot}} & \\ 1 & & 0 \end{pmatrix} \in \mathcal{U}_n .$$

Thus consider the matrices

$$(2)\qquad A = \begin{pmatrix} a_1 & & 0 \\ & \ddots & \\ 0 & & a_n \end{pmatrix} \begin{pmatrix} 1 & & a_{\mu\nu} \\ & \ddots & \\ 0 & & 1 \end{pmatrix},$$

which satisfy

(i) $a_i \in \{1, p\}$, $i = 1,\ldots,n$,

(ii) $a_1 \cdot \ldots \cdot a_n = p^j$,

(iii) $a_{\mu\nu}$ runs through a set of representatives of $\mathbb{Z}$ mod a_ν for $1 \le \mu < \nu \le n$.

Then the matrices (2) run through $p^{j(j-1)/2}$ sets of representatives of $\mathcal{U}_n \backslash T_{n,j}(p)$ due to the proof of (7.2). If $1 \le i_1 < \ldots < i_j \le n$ then

$$\iota\big(S_{n,i_1}(p) \cdot \ldots \cdot S_{n,i_j}(p)\big)$$

runs through the matrices (2) with $a_{i_\nu} = p$, $\nu = 1,\ldots,j$. Thus the assertion follows.□

Then a combination of (11.1) and (9.3) yields the decomposition of the arising polynomial.

(11.2) Corollary *If* p *is a prime,*

$$\sum_{j=0}^{n} (-1)^j\, p^{j(j-1)/2}\, \nu\big(T_{n,j}(p)\big)\, X^j$$
$$= \big(1 - S_{n,1}(p)X\big) \cdot \big(1 - S_{n,2}(p)X\big) \cdot \ldots \cdot \big(1 - S_{n,n}(p)X\big)$$

holds as a polynomial in $\mathcal{H}(\mathcal{B}_n, M_n)[X]$.

The corollary corresponds to (10.5) on the double coset level. The case $n = 2$ was already dealt with in IV(5.2).

(11.3) Exercises. a) Let $n > 1$. Show that the Hecke algebra $\mathcal{H}(\mathcal{B}_n, M_n)$ fails to be commutative and contains zero-divisors.
b) Determine the numbers of right and double cosets relative to $\mathcal{B}_n$ in the set

$$\left\{ A \in M_n \mid \det A = p^2 \right\}, \text{ whenever } p \text{ is a prime.}$$

c) Show that

$$\#\left(\mathcal{B}_n \backslash S_{n,j}(p)\right) = \#\left(S_{n,n-j}(p)/\mathcal{B}_n\right) = p^{j-1} \text{ for } 1 \le j \le n .$$

d) Recall from (5.5) that $(\mathcal{P}(m,n,\mathbb{Z}), \mathcal{P}(m,n;\mathbb{Q}))$ is a Hecke pair. Consider the "integral part" $\mathcal{M}(m,n;\mathbb{Z}) := \mathcal{P}(m,n;\mathbb{Q}) \cap \mathrm{Mat}(n;\mathbb{Z})$. Do there exist embeddings of $\mathcal{H}_{m+n}$ into the Hecke algebra $\mathcal{H}(\mathcal{P}(m,n;\mathbb{Z}), \mathcal{M}(m,n;\mathbb{Z}))$ and of the latter one into $\mathcal{H}(B_{m+n}, M_{m+n})$?

Remarks

The description of the Hecke algebra associated with the unimodular group can be found in the books of A.N. Andrianov [4] , E. Freitag [18] , and G. Shimura [58] . The Rationality Theorem (9.3) resp. (9.4) is due to T. Tamagawa [62] . But the proof presented in this chapter is different from the ones in the above mentioned works. Our presentation follows the lines given in the description of the anologous Hecke algebra over the Hurwitz order in [38] . The proof of the multiplicativity of the elementary divisor matrices using q-kernels is also modelled after the corresponding procedure for the Hurwitz order in [37] . Hecke operators for $GL(n;\mathbb{Z})$ are for instance applied by A. Terras [64] . The description of $\tilde{\mathcal{H}}_{n,p}$ in terms of symmetric polynomials is due to I. Satake [53] .

A standard reference for the elementary divisor theory including the more general results in the Exercises is M. Newman's book [49] . Our presentation is influenced by M. Koecher [32] .

Chapter VI

The Hecke Algebra Associated with the Siegel Modular Group

In this chapter we consider another generalization of chapter IV. We determine the structure of the Hecke algebra associated with the Siegel modular group of degree n. The Siegel modular group of degree 1 is the group $SL(2;\mathbb{Z})$. In this case the associated Hecke algebra was therefore already described in chapter IV. Again the Hecke algebra turns out to be a polynomial ring in infinitely many algebraically independent indeterminates. Moreover we derive the commutation relation with the Siegel Φ-operator for a set of generators of the Hecke algebra.

Throughout the whole chapter we will assume that the ground ring R is equal to $\mathbb{Z}$.

1. Symplectic Similitudes

Consider the matrix

$$J = \begin{pmatrix} 0 & I \\ -I & 0 \end{pmatrix} \in Mat(2n;\mathbb{Z}), \quad I = I^{(n)} .$$

Given $q \in \mathbb{N}$ define the set

$$\Delta_n(q) := \{ M \in Mat(2n;\mathbb{Z}) \mid M'JM = qJ \} .$$

One obviously has

(1) $$M \in \Delta_n(q),\ N \in \Delta_n(r) \Rightarrow MN \in \Delta_n(qr) .$$

Then clearly

$$\Delta_n := \bigcup_{q=1}^{\infty} \Delta_n(q)$$

becomes a submonoid of $Inv(2n;\mathbb{Z})$. The matrices in Δ_n are called *integral symplectic similitudes.* $\Gamma_n := \Delta_n(1)$ is a group and is called the *Siegel modular group of degree* n. Calculation of the determinant immediately leads to

$$\Gamma_n \subset \mathcal{U}_{2n} .$$

Given a matrix $M \in \Delta_n$ we will always assume that M has the standard form

$$M = \begin{pmatrix} A & B \\ C & D \end{pmatrix}, \quad A,B,C,D \in \mathrm{Mat}(n;\mathbb{Z}) .$$

At first we characterize the equation $M'JM = qJ$. A simple verification leads to

(1.1) Lemma. *Let* $q \in \mathbb{N}$ and $M = \begin{pmatrix} A & B \\ C & D \end{pmatrix} \in \mathrm{Mat}(2n;\mathbb{Z})$. *Then the following assertions are equivalent:*

(i) $M \in \Delta_n(q)$.

(ii) $M' \in \Delta_n(q)$.

(iii) $M \in \mathrm{Inv}(2n;\mathbb{Z})$ *and* $qM^{-1} \in \Delta_n(q)$.

(iv) $M \in \mathrm{Inv}(2n;\mathbb{Z})$ *and*

$$M^{-1} = \frac{1}{q}\begin{pmatrix} D' & -B' \\ -C' & A' \end{pmatrix} .$$

(v) $A'C = C'A$, $B'D = D'B$, $A'D - C'B = qI$.

(vi) $AB' = BA'$, $CD' = DC'$, $AD' - BC' = qI$.

It follows from the lemma that the modular group Γ in chapter IV coincides with the Siegel modular group Γ_1 of degree 1 and the monoid $M = \mathrm{Inv}^+(2;\mathbb{Z})$ with Δ_1 .

The group generated by Δ_n is given by

$$\begin{aligned} \bar{\Delta}_n &= \left\{ \tfrac{1}{q} M \mid q \in \mathbb{N},\ M \in \Delta_n \right\} \\ &= \left\{ M \in \mathrm{Mat}(2n;\mathbb{Q}) \mid M'JM = \nu J \ \text{ for some positive } \nu \in \mathbb{Q} \right\}, \end{aligned}$$

which is the group of *rational symplectic similitudes* and contains the rational symplectic group $Sp(n;\mathbb{Q})$ as a normal subgroup.

Next we describe some parabolic subgroups in $\bar{\Delta}_n$. Given $0 \le j < n$ consider the subgroup

$$\bar{\Delta}_{n,j} = \bar{\Delta}_n \cap \mathcal{P}(n+j,n-j;\mathbb{Q}) = \left\{ M = \begin{pmatrix} * & * \\ 0 & * \end{pmatrix} \in \bar{\Delta}_n \mid 0 = 0^{(n-j,n+j)} \right\}$$

of $\bar{\Delta}_n$. We use the obvious abbreviations

$$\Delta_{n,j} = \bar{\Delta}_{n,j} \cap \Delta_n , \quad \Gamma_{n,j} = \bar{\Delta}_{n,j} \cap \Gamma_n .$$

(1.2) Lemma. *Let* $q \in \mathbb{N}$, $0 < j < n$ *and* $M = \begin{pmatrix} A & B \\ C & D \end{pmatrix} \in \Delta_n(q) \cap \Delta_{n,j}$. *Then one has*

$$A = \begin{pmatrix} A_1 & 0 \\ A_3 & A_4 \end{pmatrix}, \quad B = \begin{pmatrix} B_1 & B_2 \\ B_3 & B_4 \end{pmatrix}, \quad C = \begin{pmatrix} C_1 & 0 \\ 0 & 0 \end{pmatrix}, \quad D = \begin{pmatrix} D_1 & D_2 \\ 0 & D_4 \end{pmatrix},$$

where the blocks with index 1 *are* $j \times j$ *matrices. Moreover*

$$M_1 := \begin{pmatrix} A_1 & B_1 \\ C_1 & D_1 \end{pmatrix} \in \Delta_j(q)$$

holds and the induced mapping

$$\bar\Delta_{n,j} \longrightarrow \bar\Delta_j\,,$$
$$M \longmapsto M_1\,,$$

is a surjective homomorphism of the groups.

Proof. $M \in \Delta_{n,j}$ implies block decompositions

$$A = \begin{pmatrix} A_1 & A_2 \\ A_3 & A_4 \end{pmatrix},\quad B = \begin{pmatrix} B_1 & B_2 \\ B_3 & B_4 \end{pmatrix},\quad C = \begin{pmatrix} C_1 & C_2 \\ 0 & 0 \end{pmatrix},\quad D = \begin{pmatrix} D_1 & D_2 \\ 0 & D_4 \end{pmatrix}.$$

Because of $qM^{-1} \in \Delta_n(q) \cap \Delta_{n,j}$ we conclude

$$A_2 = C_2 = 0 \quad \text{and} \quad M_1 \in \Delta_j(q)$$

from (1.1iv) . □

Analogous arguments can of course be applied to elements of $\Delta_n \cap \mathcal{P}(n-j,n+j;\mathbb{Q})$. The result is given in

(1.3) Lemma. *Let* $q \in \mathbb{N}$, $0 < j < n$ *and* $M = \begin{pmatrix} A & B \\ C & D \end{pmatrix} \in \Delta_n(q) \cap \mathcal{P}(n-j,n+j;\mathbb{Q})$. *Then one has*

$$A = \begin{pmatrix} A_1 & A_2 \\ 0 & A_4 \end{pmatrix},\quad B = \begin{pmatrix} B_1 & B_2 \\ B_3 & B_4 \end{pmatrix},\quad C = \begin{pmatrix} 0 & 0 \\ 0 & C_4 \end{pmatrix},\quad D = \begin{pmatrix} D_1 & 0 \\ D_3 & D_4 \end{pmatrix},$$

where the blocks with index 4 *are* $j \times j$ *matrices. Moreover*

$$M_4 := \begin{pmatrix} A_4 & B_4 \\ C_4 & D_4 \end{pmatrix} \in \Delta_j(q)$$

holds.

The case $j = 0$ is dealt with separately. If we use $\Gamma_n \subset \mathcal{U}_{2n}$ and apply (1.1) , we obtain

(1.4) Lemma. $\Gamma_{n,0} = \left\{\begin{pmatrix} A & B \\ C & D \end{pmatrix} \in \Gamma_n \mid C = 0\right\} = \left\{\begin{pmatrix} U' & 0 \\ 0 & U^{-1} \end{pmatrix}\begin{pmatrix} I & S \\ 0 & I \end{pmatrix} \mid U \in \mathcal{U}_n,\ S \in Sym(n;\mathbb{Z})\right\}$

$$= \left\{\begin{pmatrix} I & T \\ 0 & I \end{pmatrix}\begin{pmatrix} U' & 0 \\ 0 & U^{-1} \end{pmatrix} \mid U \in \mathcal{U}_n,\ T \in Sym(n;\mathbb{Z})\right\}.$$

Finally we introduce the usual embedding of symplectic groups of different degrees. Given $q \in \mathbb{N}$ and $0 < j < n$ define the mapping

$$\Delta_j(q) \times \Delta_{n-j}(q) \longrightarrow \Delta_n(q)\,,$$
$$(M_1,M_2) \longmapsto M_1 \times M_2\,,$$

$$\begin{pmatrix} A_1 & B_1 \\ C_1 & D_1 \end{pmatrix} \times \begin{pmatrix} A_2 & B_2 \\ C_2 & D_2 \end{pmatrix} := \begin{pmatrix} A_1 & 0 & B_1 & 0 \\ 0 & A_2 & 0 & B_2 \\ C_1 & 0 & D_1 & 0 \\ 0 & C_2 & 0 & D_2 \end{pmatrix}.$$

Under obvious conditions on the size we obtain

$$(M_1 \times M_2) \cdot (N_1 \times N_2) = (M_1N_1) \times (M_2N_2)\,. \tag{2}$$

In the Exercises we will describe the Hecke algebras of two other groups, which are related with the Siegel modular group. Therefore let $\mathfrak{o}$ denote the ring of integers in an imaginary quadratic number field and set

$$\Delta_n^{\mathfrak{o}}(q) \ := \ \{ M \in \mathrm{Mat}(2n;\mathfrak{o}) \mid \overline{M}'JM = qJ \} \quad \text{for} \quad q \in \mathbb{N} .$$

Then $\Gamma_n^{\mathfrak{o}} := \Delta_n^{\mathfrak{o}}(1)$ is a group, which is called the *Hermitian modular group of degree* n *associated with* $\mathfrak{o}$. This group was extensively studied by H. Braun [10] in analogy with the Siegel modular group. We will use the analogous notations with a superscript $\mathfrak{o}$ for this case throughout the whole chapter. Particular difficulties arise in the case, where the class number of $\mathfrak{o}$ is greater than 1.

Moreover consider the symmetric matrix

$$Q \ = \ \begin{pmatrix} 0 & I \\ I & 0 \end{pmatrix} \in \mathrm{Sym}(2n;\mathbb{Z}) \ , \ \text{where} \ \ I = I^{(n)} .$$

Given $q \in \mathbb{N}$ set

$$\Delta_n^{*}(q) \ := \ \{ M \in \mathrm{Mat}(2n;\mathbb{Z}) \mid M'QM = qQ \} .$$

Then $\Gamma_n^{*} := \Delta_n^{*}(1)$ again turns out to be a group, the so-called *modified modular group of degree* n (cf. [39]), which in fact is a split orthogonal group. The corresponding notations with a superscript * will refer to this case throughout the whole chapter.

(1.5) Exercises. a) Given $\ell \in \mathbb{N}$ show that

$$\Gamma_n(\ell) := \{ M \in \Gamma_n \mid M \equiv I \mod \ell \} \ \text{ and } \ \Gamma_{n,0}(\ell) \ := \ \{ M \in \Gamma_n \mid C \equiv 0 \mod \ell \}$$

are subgroups of Γ_n with finite index. Compute the index in each case (Hint: [31]).
b) Given $q \in \mathbb{N}$ and $M \in \Delta_n(q)$ show that $\# \ker(M|q) = q^n$ (cf. V.4) holds.
c) Verify the analogous results for $\Gamma_n^{\mathfrak{o}}$. Show that

$$\Gamma_1^{\mathfrak{o}} = \{ \varepsilon M \mid M \in SL(2;\mathbb{Z}) \ , \ \varepsilon \ \text{ unit in } \mathfrak{o} \} \ , \ \Delta_1^{\mathfrak{o}} = \{ aM \mid M \in \mathrm{Inv}^{+}(2;\mathbb{Z}) \ , \ 0 \neq a \in \mathfrak{o} \} .$$

d) Derive the analogous results for Γ_n^{*}. Show that

$$\Gamma_1^{*} = \{ \pm I, \pm Q \} \ , \quad \Delta_1^{*} = \left\{ \begin{pmatrix} a & 0 \\ 0 & d \end{pmatrix} \mid a,d \in \mathbb{Z}\backslash\{0\} \right\} \cup \left\{ \begin{pmatrix} 0 & b \\ c & 0 \end{pmatrix} \mid b,d \in \mathbb{Z}\backslash\{0\} \right\} .$$

2. The Symplectic Elementary Divisor Theorem

In this section we prove the analogue of V.2 for the Siegel modular group in place of the unimodular group as well as Δ_n in place of $\mathrm{Inv}(n;\mathbb{Z})$.

(2.1) Theorem. *The Siegel modular group* Γ_n *is generated by the matrices*

(1) $$J = \begin{pmatrix} 0 & I \\ -I & 0 \end{pmatrix} , \ \begin{pmatrix} U' & 0 \\ 0 & U^{-1} \end{pmatrix} , \ U \in \mathcal{U}_n \ , \ \begin{pmatrix} I & S \\ 0 & I \end{pmatrix} , \ S \in \mathrm{Sym}(n;\mathbb{Z}) .$$

(2.2) Theorem. *Symplectic Elementary Divisor Theorem*
Given $M \in \Delta_n(q)$ *there exist* $K, L \in \Gamma_n$ *as well as uniquely determined positive integers* $a_1,\ldots,a_n,d_1,\ldots,d_n \in \mathbb{N}$ *such that* $a_i d_i = q$, $i = 1,\ldots,n$, *and*

$$KML = \operatorname{diag}\left(a_1,\ldots,a_n,d_1,\ldots,d_n\right), \quad a_1 \mid a_2 \mid \ldots \mid a_n \mid d_n \mid d_{n-1} \mid \ldots \mid d_1 .$$

Proof of (2.1) and (2.2). The matrices (1) belong to Γ_n according to (1.1) . Let $G(n)$ denote the subgroup of Γ_n generated by the matrices (1) . Because of $\Gamma_1 = SL(2;\mathbb{Z})$ we can apply V(2.3) in order to conclude

$$G(1) = \Gamma_1 = SL(2;\mathbb{Z}) . \tag{2}$$

We also get

$$\begin{pmatrix} I & 0 \\ S & I \end{pmatrix} = J \begin{pmatrix} I & -S \\ 0 & I \end{pmatrix} J^{-1} \in G(n) \quad \text{for } S \in Sym(n;\mathbb{Z}) . \tag{3}$$

Claim 1. *Let* $n > 1$ *and* $M \in G(n-1)$, *then* $I^{(2)} \times M \in G(n)$.

Proof. One easily checks

$$\left(\begin{pmatrix} 1 & 0 \\ 1 & 1 \end{pmatrix} \times I \right) \left(\begin{pmatrix} 1 & -1 \\ 0 & 1 \end{pmatrix} \times I \right) J \left(\begin{pmatrix} 1 & -1 \\ 0 & 1 \end{pmatrix} \times I \right) = I^{(2)} \times J^{(2n-2)} .$$

The remaining part is obvious. □

The following statement is proved by induction on n :

$\mathcal{S}(n)$: Given $M \in \Delta_n(q)$ there exist $K, L \in G(n)$ as well as $a_1,\ldots,a_n,d_1,\ldots,d_n \in \mathbb{N}$ such that $a_i d_i = q$, $i = 1,\ldots,n$, and

$$KML = \operatorname{diag}\left(a_1,\ldots,a_n,d_1,\ldots,d_n\right), \quad a_1 \mid a_2 \mid \ldots \mid a_n \mid d_n \mid d_{n-1} \mid \ldots \mid d_1 .$$

Let $\mathcal{M}$ denote the set of positive integers, which occur as entries of the matrices in $G(n)MG(n)$. Then $\mathcal{M}$ is non-empty due to $-I \in G(n)$. Choose

$$a := \min \mathcal{M} .$$

Without restriction assume that a occurs as an entry of M . If we replace M by a product of M and J resp. J^{-1} , we may assume that a occurs in the A-block of M . Then multiply by $\begin{pmatrix} P & 0 \\ 0 & P \end{pmatrix}$, where P is a suitable permutation matrix, and assume $m_{11} = a$.

Claim 2. *If* $\bar{M} \in G(n)MG(n)$ *with* $\bar{m}_{11} = a$, *then* a *divides the entries in the first column and the first row of* $\bar{M}$.

Proof. Given $2 \le j \le 2n$ choose $x_j \in \mathbb{Z}$ with $0 \le \bar{m}_{1j} + x_j a < a$. If $2 \le j \le n$ multiply by $\begin{pmatrix} U' & 0 \\ 0 & U^{-1} \end{pmatrix}$, $U = I + x_j I_{j1}$, from the right hand side, if $j = n+1$ by $\begin{pmatrix} 1 & x_{n+1} \\ 0 & 1 \end{pmatrix} \times I$ and if $n+1 < j \le 2n$ by $\begin{pmatrix} I & S \\ 0 & I \end{pmatrix}$, $S = x_j\left(I_{1,j-n} + I_{j-n,1}\right)$, from the right hand side. We conclude

$$\bar{m}_{1j} + x_j a \in \mathcal{M} \cup \{0\}$$

and $\bar{m}_{1j} = -x_j a$ from the choice of a. Apply analogous arguments to the first column of $\bar{M}$. □

Following the proof of Claim 2 we get

Claim 3. *There exist* $d \in \mathbb{N}$ *with* $a \mid d$ *and* $ad = q$ *as well as* $M_1 \in \Delta_{n-1}(q)$ *such that*

$$\begin{pmatrix} a & 0 \\ 0 & d \end{pmatrix} \times M_1 \in G(n) M G(n) .$$

Proof. Following the proof of Claim 2 we find a matrix $\bar{M}$ in $G(n)MG(n)$ with ae_1 as its first column and with ae_1' as its first row. Application of (1.3) with $j = n-1$ to M as well as to M' yields $d \in \mathbb{N}$ and $M_1 \in \Delta_{n-1}(q)$ such that $ad = q$ and

$$\begin{pmatrix} a & 0 \\ 0 & d \end{pmatrix} \times M_1 \in G(n) M G(n) .$$

Now $a \mid d$ follows from Claim 2 and 1(2) in the identity

$$\left(\begin{pmatrix} 1 & 1 \\ 0 & 1 \end{pmatrix} \times I\right) \left(\begin{pmatrix} a & 0 \\ 0 & d \end{pmatrix} \times M_1\right) = \begin{pmatrix} a & d \\ 0 & d \end{pmatrix} \times M_1 .$$ □

If $n = 1$, the statement is proved. Otherwise Claim 1 and the induction hypothesis yield a representative

$$\operatorname{diag}(a, a_2, \ldots, a_n, d, d_2, \ldots, d_n), \quad a_2 \mid \ldots \mid a_n \mid d_n \mid d_{n-1} \mid \ldots \mid d_2$$

in $G(n)MG(n)$. Multiplication of this matrix by $\begin{pmatrix} U' & 0 \\ 0 & U^{-1} \end{pmatrix}$ from the left hand side, where $U = I + I_{21}$, and Claim 2 imply $a \mid a_2$.

The elementary divisor matrix of M equals $\operatorname{diag}(a_1, \ldots, a_n, d_n, \ldots, d_1)$. Hence the uniqueness follows from V(3.3) and (2.2) is proved.

Given $M \in \Gamma_n$ choose $q = 1$ and apply $\mathcal{S}(n)$. The resulting diagonal matrix must be $I^{(2n)}$ because of $M \in \mathcal{U}_{2n}$. Hence $M \in G(n)$ and $G(n) = \Gamma_n$ follow. □□

The fact that the elementary divisior matrix of M in (2.2) equals

$$\operatorname{diag}(a_1, \ldots, a_n, d_n, \ldots, d_1)$$

is used in order to obtain

(2.3) Corollary. *Let* $M, N \in \Delta_n(q)$. *Then one has*

$$\Gamma_n M \Gamma_n = \Gamma_n N \Gamma_n \quad \textit{if and only if} \quad \mathcal{U}_{2n} M \mathcal{U}_{2n} = \mathcal{U}_{2n} N \mathcal{U}_{2n} .$$

(2.4) Remark. a) H. Maaß [45] showed that Γ_n is already generated by

$$J \text{ and } \begin{pmatrix} I & S \\ 0 & I \end{pmatrix}, \; S \in \operatorname{Sym}(n; \mathbb{Z}) .$$

Therefore the identity

$$\begin{pmatrix} 2 & 1 \\ 1 & 1 \end{pmatrix} \begin{pmatrix} 0 & 1 \\ 1 & -1 \end{pmatrix} \begin{pmatrix} 0 & 1 \\ 1 & 0 \end{pmatrix} = \begin{pmatrix} 1 & 1 \\ 0 & 1 \end{pmatrix}$$

and V(2.1) yield that $\mathcal{U}_n$ is generated by $\mathcal{U}_n \cap \mathrm{Sym}(n;\mathbb{Z})$. Given $U \in \mathcal{U}_n \cap \mathrm{Sym}(n;\mathbb{Z})$ he calculates

$$\begin{pmatrix} I & U \\ 0 & I \end{pmatrix} \begin{pmatrix} 0 & I \\ -I & 0 \end{pmatrix} \begin{pmatrix} I & U^{-1} \\ 0 & I \end{pmatrix} \begin{pmatrix} 0 & I \\ -I & 0 \end{pmatrix} \begin{pmatrix} I & U \\ 0 & I \end{pmatrix} \begin{pmatrix} 0 & I \\ -I & 0 \end{pmatrix} = \begin{pmatrix} U & 0 \\ 0 & U^{-1} \end{pmatrix} .$$

b) L.K. Hua and I. Reiner [22] showed that Γ_n , $n \geq 2$, is always generated by the following four matrices

$$J^{(2)} \times I , \ \begin{pmatrix} 1 & 1 \\ 0 & 1 \end{pmatrix} \times I , \ \begin{pmatrix} U' & 0 \\ 0 & U^{-1} \end{pmatrix} , \ U = I + I_{12} , \ P_\pi , \ \pi = \tau_{1,2} \circ \tau_{2,3} \circ \ldots \circ \tau_{n-1,n} \, .$$

(2.5) Exercises. a) Show that $\det M = 1$ holds for all $M \in \Gamma_n$.
b) Let $\ell, q \in \mathbb{N}$ and recall the definition of $\Gamma_{n,0}(\ell)$ from (1.5) . Show that the double coset $\Gamma_{n,0}(\ell) M \Gamma_{n,0}(\ell)$, where $M \in \Delta_n(q)$ with $C \equiv 0 \mod \ell$, contains a unique representative of the normal form in (2.2) , whenever q and ℓ are relatively prime.
c) Let $\ell > 1$ and $n > 1$. Show that the subgroup $\Gamma_{n,0}(\ell)$ is generated by the matrices

$$M \times I^{(2n-2)} , \ M \in \Gamma_{1,0}(\ell) , \ \begin{pmatrix} U' & 0 \\ 0 & U^{-1} \end{pmatrix} , \ U \in \mathcal{U}_n , \ \begin{pmatrix} I & S \\ 0 & I \end{pmatrix} , \ \begin{pmatrix} I & 0 \\ \ell T & I \end{pmatrix} , \ S, T \in \mathrm{Sym}(n;\mathbb{Z}) .$$

d) Demonstrate the analogue of (2.1) , (2.2) , (2.3) for $\Gamma_n^{\mathfrak{o}}$, whenever $\mathfrak{o}$ has the class number 1 , i.e., $\mathfrak{o}$ is a p.i.d. . Confer [30] in order to derive an analogue of (2.1) for an $\mathfrak{o}$ with arbitrary class number.
e) Let $n > 1$. Show that Γ_n^* is generated by the matrices

$$Q^{(2)} \times I , \ \begin{pmatrix} U' & 0 \\ 0 & U^{-1} \end{pmatrix} , \ U \in \mathcal{U}_n , \ \begin{pmatrix} I & S \\ 0 & I \end{pmatrix} , \ S \in \mathrm{Mat}(n;\mathbb{Z}) , \ S' = -S .$$

Derive the analogue of (2.2) and (2.3) for Γ_n^* .
f) Given $M = \begin{pmatrix} A & B \\ C & D \end{pmatrix} \in \Gamma_n^*$ show that

$$\det M = (-1)^r , \text{ where } r = \mathrm{rank}\, C .$$

3. Right Coset Decompositions

In this section we prove the analogue of section V.5 for the Siegel modular group. Representatives of the right cosets may be chosen as upper "triangular" matrices, which means that the C-block is equal to 0 .

(3.1) Lemma. *Given* $0 \neq g \in \mathbb{Z}^{2n}$ *there exists* $M \in \Gamma_n$ *such that*

$$Mg = (\delta, 0, \ldots, 0)' , \ \delta = \text{g.c.d. } (g) .$$

Proof. First of all replace g by $\begin{pmatrix} U' & 0 \\ 0 & U^{-1} \end{pmatrix} g$, $U \in \mathcal{U}_n$, and assume

$$g = (a_1,\dots,a_n,c,0,\dots,0)'$$

according V(3.4) without restriction. Then apply V(3.4) or IV(1.1) to the column $\binom{a_1}{c}$ and determine $M_1 \in \mathcal{SU}_2 = \Gamma_1$ such that $M_1\binom{a_1}{c} = \binom{*}{0}$. Finally choose $V \in \mathcal{U}_n$ in view of V(3.4) such that

$$\begin{pmatrix} V' & 0 \\ 0 & V^{-1} \end{pmatrix} (M_1 \times I) g = (\delta,0,\dots,0)'$$

holds with some $\delta \in \mathbb{N}$. Hence $\delta = \text{g.c.d.}(g)$ follows from $\Gamma_n \subset \mathcal{U}_{2n}$. □

This has an immediate consequence for representatives of right cosets.

(3.2) Lemma. *Given* $M \in \Delta_n$ *the right coset* $\Gamma_n M$ *contains a representatives, which is contained in* $\Delta_{n,0}$, *i.e., which has the form*

$$\begin{pmatrix} A & B \\ 0 & D \end{pmatrix}.$$

Proof. We use induction on n, where $n = 1$ follows from (3.1) or IV(1.2). If $n > 1$ we apply (3.1) to the first column of M. Then we can apply (1.3) to $j = n-1$. In the notation of (1.3) we have $M_4 \in \Delta_{n-1}$. Due to the induction hypothesis we can choose $K \in \Gamma_{n-1}$ such that KM_4 has the desired form. Then the C-block of the matrix

$$\left(I^{(2)} \times K\right) M$$

is equal to 0. □

Given $M \in \Delta_n(q)$ we have a look at the set $\Gamma_n M \Gamma_n \cap \Delta_{n,0}$, which is not empty due to (3.2). Then set

$$\mathcal{D}(\Gamma_n M \Gamma_n) = \left\{ D \in \text{Mat}(n;\mathbb{Z}) \mid \text{there exists } \begin{pmatrix} A & B \\ 0 & D \end{pmatrix} \text{ in } \Gamma_n M \Gamma_n \right\}.$$

Note that

(1) $$\mathcal{D}(\Gamma_n M \Gamma_n) \subset \left\{ G \in \text{Inv}(n;\mathbb{Z}) \mid qG^{-1} \in \text{Mat}(n;\mathbb{Z}) \right\}$$

holds due to (1.1) and that $\mathcal{D}(\Gamma_n M \Gamma_n)$ decomposes into full double cosets relative to $\mathcal{U}_n$. Whenever $D \in \mathcal{D}(\Gamma_n M \Gamma_n)$ is given, set

$$\mathcal{B}(D,\Gamma_n M \Gamma_n) = \left\{ B \in \text{Mat}(n;\mathbb{Z}) \mid \text{there exists } \begin{pmatrix} A & B \\ 0 & D \end{pmatrix} \text{ in } \Gamma_n M \Gamma_n \right\}$$

Note that (1.1) implies

(2) $$BD^{-1} \in \text{Sym}(n;\mathbb{Q}) \quad \text{for } \begin{pmatrix} A & B \\ 0 & D \end{pmatrix} \in \Delta_n(q).$$

Thus we can define an equivalence relation on $\mathcal{B}(D,\Gamma_n M \Gamma_n)$ via

(3) $$B \equiv B^* \bmod D \iff (B - B^*)D^{-1} \in \text{Sym}(n;\mathbb{Z}).$$

Note that

$$(4) \qquad B \equiv B^* \mod D \iff \Gamma_{n,o}\begin{pmatrix} A & B \\ 0 & D \end{pmatrix} = \Gamma_{n,o}\begin{pmatrix} A & B^* \\ 0 & D \end{pmatrix}$$

holds due to (1.4) , where $A = qD'^{-1}$.

(3.3) Lemma. *Let* $M \in \Delta_n(q)$, $D \in \mathcal{D}(\Gamma_n M \Gamma_n)$.

a)
$$\#\left(\mathcal{B}(D,\Gamma_n M\Gamma_n) \mod D\right) = \#\left(\mathcal{B}(UDV, \Gamma_n M\Gamma_n) \mod UDV\right)$$

holds for all $U,V \in \mathcal{U}_n$.

b) *If* $\operatorname{diag}(d_1,\ldots,d_n)$ *is the elementary divisor matrix of* D , *one has*

$$\#\left(\mathcal{B}(D,\Gamma_n M\Gamma_n) \mod D\right) \leq d_1^{\,n}\, d_2^{\,n-1}\cdot\ldots\cdot d_n .$$

Proof. a) One has

$$\mathcal{B}(UDV,\Gamma_n M\Gamma_n) = U'^{-1}\mathcal{B}(D,\Gamma_n M\Gamma_n)\, V .$$

b) Assume $D = \operatorname{diag}(d_1,\ldots,d_n)$ due to a) and V(2.2) . Then (2) and $d_1 | \ldots | d_n$ lead to

$$(5) \qquad \mathcal{B}(D,\Gamma_n M\Gamma_n) \subset \left\{ B=(b_{jk}) \mid b_{jk}\in \mathbb{Z} \text{ for } 1 \leq k \leq j \leq n ,\ b_{jk} = b_{kj}d_k d_j^{-1} \text{ for } j < k \right\} .$$

According to the equivalence relation (3) the elements b_{jk} may be reduced mod d_k for $1 \leq k \leq j \leq n$. Thus the right hand side of (5) consists exactly of $d_1^{\,n} d_2^{\,n-1}\cdot\ldots\cdot d_n$ equivalence classes mod D . □

Now we derive representatives of the right cosets.

(3.4) Theorem. a) *Let* $M \in \Delta_n(q)$. *Then a set of representatives of the right cosets relative to* Γ_n *in* $\Gamma_n M\Gamma_n$ *is given by the matrices*

$$N = \begin{pmatrix} A & B \\ 0 & D \end{pmatrix}, \quad A = qD'^{-1} ,$$

where

(i) D *runs through a set of representatives of* $\mathcal{U}_n \backslash \mathcal{D}(\Gamma_n M\Gamma_n)$,

(ii) B *runs through a set of representatives of* mod D *incongruent matrices in* $\mathcal{B}(D,\Gamma_n M\Gamma_n)$, *whenever* D *is given.*

b) $\Delta_n(q)$ *decomposes into finitely many right cosets relative to* Γ_n . *In particular* $(\Gamma_n,\bar{\Delta}_n)$ *is a Hecke pair.*

Proof. a) It follows from (3.2) , (1.4) , (4) , and the considerations above that each right coset contains a representative described by (i) and (ii) . If representatives

$$N = \begin{pmatrix} A & B \\ 0 & D \end{pmatrix}, \quad N^* = \begin{pmatrix} A^* & B^* \\ 0 & D^* \end{pmatrix} \text{ with } \Gamma_n N = \Gamma_n N^*$$

are given, we conclude

$$N^* N^{-1} \in \Gamma_n \cap \bar{\Delta}_{n,0} = \Gamma_{n,0} .$$

Now (1.4) leads to $D^* = UD$ for some $U \in \mathcal{U}_n$. Then (i) implies $U = I$ and $D = D^*$. Next (1.4) leads to $B^* = B + SD$ for some $S \in Sym(n;\mathbb{Z})$. Then (ii) implies $S = 0$ and $B = B^*$.

b) The set

$$\{G \in Inv(n;\mathbb{Z}) \mid q\,G^{-1} \in Mat(n;\mathbb{Z})\}$$

decomposes into finitely many right cosets relative to $\mathcal{U}_n$. Then a) and (3.3) complete the proof.

Given $M \in \bar{\Delta}$ choose $\ell \in \mathbb{N}$ with $\ell M \in \Delta_n$ and note that

$$\#\left(\Gamma_n \backslash \Gamma_n M \Gamma_n\right) = \#\left(\Gamma_n \backslash \Gamma_n (\ell M) \Gamma_n\right) . \qquad \square$$

According to V(5.2) the representatives D in (i) can be chosen as upper triangular matrices. Then (3.4) in particular yields

(3.5) Corollary. *Given* $M \in \Delta_n$ *the right coset* $\Gamma_n M$ *possesses a representative, which is contained in*

$$\bigcap_{j=0}^{n-1} \Delta_{n,j} .$$

(3.6) Exercises. a) Given $M \in \Delta_n(q)$ show that $\Gamma_n \cap M^{-1}\Gamma_n M$ is a subgroup of Γ_n, which contains $\Gamma_n(q)$ and therefore has finite index in Γ_n (cf. (1.5)).

b) Determine the number of right cosets relative to Γ_n in $\Delta_n(p)$, whenever p is a prime.

c) Find an example of a double coset, where the inclusion (5) is proper.

d) Obtain the analogous descriptions for the right cosets relative to $\Gamma_n^{\mathfrak{o}}$ in $\Delta_n^{\mathfrak{o}}$ provided that $\mathfrak{o}$ has the class number 1 .

e) Derive the analogue in the case Γ_n^*, $n > 1$. Give an explicit description for $n = 1$.

4. The Hecke Algebra Associated with (Γ_n, Δ_n)

We investigate the Hecke algebra

$$\mathcal{H}_n := \mathcal{H}(\Gamma_n, \Delta_n) \quad \text{over} \quad R := \mathbb{Z} .$$

The proofs are similar to the ones in section V.6 .

One has $\det M = q^n$ for $M \in \Delta_n(q)$ and for $M, N \in \Delta_n$

$$\Gamma_n K = \Gamma_n L \iff \mathcal{U}_{2n} K = \mathcal{U}_{2n} L .$$

In particular the same proof as for V(6.1) now gives

(4.1) Lemma. a) $\mathcal{H}_n$ *is a commutative algebra.*
b) *Given* $M \in \Delta_n$ *and* $0 \neq \lambda \in \mathbb{Z}$ *one has*

$$(\Gamma_n M \Gamma_n) \cdot (\Gamma_n(\lambda I)\Gamma_n) = \Gamma_n(\lambda M)\Gamma_n .$$

c) *Given* $M \in \Delta_n(q)$ *and* $N \in \Delta_n(r)$, *where* q,r *are relatively prime, one has*

$$(\Gamma_n M \Gamma_n) \cdot (\Gamma_n N \Gamma_n) = \Gamma_n(MN)\Gamma_n .$$

It also follows from I(7.2) that the numbers of left and right cosets in each double coset relative to Γ_n coincide.

Given $q \in \mathbb{N}$ set

$$\mathcal{T}_n(q) := \sum_{M:\Gamma_n\backslash\Delta_n(q)/\Gamma_n} \Gamma_n M \Gamma_n .$$

In analogy with V(6.2) we obtain

(4.2) Corollary. *Given relatively prime* $q,r \in \mathbb{N}$ *one has*

$$\mathcal{T}_n(q) \cdot \mathcal{T}_n(r) = \mathcal{T}_n(qr) .$$

Again we decompose the Hecke algebra into its primary components. Given a prime number p consider the submonoid

$$\Delta_{n,p} := \bigcup_{\ell=0}^{\infty} \Delta_n(p^\ell)$$

of Δ_n and the subalgebra

$$\mathcal{H}_{n,p} := \mathcal{H}(\Gamma_n, \Delta_{n,p}) .$$

$\mathcal{H}_{n,p}$ is called the p-*primary component* of $\mathcal{H}_n$.

In view of the Symplectic Elementary Divisor Theorem (2.2) we can apply the same arguments as for V(6.4) in order to decompose the Hecke algebra into its primary components. Hence we get

(4.3) Theorem. *The Hecke algebra* $\mathcal{H}_n$ *coincides with the tensor product of its primary components, i.e.,*

$$\mathcal{H}_n = \bigotimes_{p \text{ prime}} \mathcal{H}_{n,p} .$$

Just as in the case of the unimodular group it suffices to investigate the primary components.

(4.4) Exercises. a) Given $M \in \Delta_n$ set

$$\nu(M) := [\Gamma_n : \Gamma_n \cap M^{-1}\Gamma_n M]$$

(cf. (3.6)) . If $K,L \in \Gamma_n$ and $M \in \Delta_n(q)$, $N \in \Delta_n(r)$, where q,r are relatively prime, show that

$$\nu(KML) = \nu(M) \text{ and } \nu(MN) = \nu(M)\,\nu(N) .$$

b) Define the arithmetical function

$$d_n(q) := \#\left(\Gamma_n\backslash\Delta_n(q)\right) \text{ for } q \in \mathbb{N} .$$

Show that d_n is multiplicative, i.e.,

$$d_n(qr) = d_n(q)\,d_n(r) \text{ for relatively prime } q,r \in \mathbb{N} .$$

c) Let $\ell > 1$ and recall the definition of $\Gamma_{n,0}(\ell)$ from (1.5) . Show that $(\Gamma_{n,0}(\ell),\bar{\Delta}_n)$ is a Hecke pair and that the attached Hecke algebra fails to be commutative.

d) Derive the analogue of (4.1) , (4.2) , (4.3) for $\mathcal{H}_n^{o}$, whenever o has the class number 1 .

e) Demonstrate the analogue for $\mathcal{H}_n^*$ in the case $n > 1$. Show that $\mathcal{H}_1^*$ is a commutative algebra but that (4.1c) does not come true in $\mathcal{H}_1^*$.

5. Particular Double Cosets in the Primary Component

As in the case of the unimodular group we investigate the primary components. Therefore fix a prime number p . Due to the Symplectic Elementary Divisor Theorem (2.2) each double coset $\Gamma_n M \Gamma_n$, $M \in \Delta_n(p^\ell)$, possesses a unique representative of the form

(1) $$M = \operatorname{diag}\left(p^{k_1},\dots,p^{k_n},p^{\ell-k_1},\dots,p^{\ell-k_n}\right) , \quad 0 \le k_1 \le\dots\le k_n \le \ell/2 .$$

Hence $\Delta_n(p)$ consists of a single double coset

$$T_n(p) = \Delta_n(p) = \Gamma_n\begin{pmatrix} I & 0 \\ 0 & pI\end{pmatrix}\Gamma_n .$$

Using the results of section V.7 one has

(2) $$\mathcal{D}(T_n(p)) = \bigcup_{j=0}^{n} T_{n,j}(p) .$$

If $D = \begin{pmatrix} I^{(n-j)} & 0 \\ 0 & pI^{(j)}\end{pmatrix}$, $0 \le j \le n$, we obtain

$$\mathcal{B}(D, T_n(p)) = \left\{\begin{pmatrix} B_1 & pB_3' \\ B_3 & B_4\end{pmatrix} \mid B_1 \in \operatorname{Sym}(n-j;\mathbb{Z}),\ B_4 \in \operatorname{Sym}(j;\mathbb{Z}) ,\ B_3 \in \operatorname{Mat}(j,n-j;\mathbb{Z})\right\}$$

and a set of representatives of $\mathcal{B}(D,T_n(p))$ mod D is given by

(3) $$\begin{pmatrix} 0 & 0 \\ 0 & S\end{pmatrix} , \quad S \in \operatorname{Sym}(j;\mathbb{Z}) \text{ mod } p .$$

Thus one has

$$\#\left(\mathcal{B}(D,T_n(p)) \text{ mod } D\right) = p^{j(j+1)/2} \quad \text{for } D \in T_{n,j}(p) ,\ 0 \le j \le n .$$

Using (2) , (3.3) , V(7.2) , and

$$\sum_{j=0}^{n} E_j(p,p^2,\dots,p^n) = \prod_{j=1}^{n} (p^j+1) ,$$

we get

(5.1) Lemma. ind $T_n(p) = \prod_{j=1}^{n} (p^j+1)$.

Next we are going to determine $T_n(p)^2$. Therefore we have to describe all the double cosets in $\Delta_n(p^2)$. Given $0 \le j \le n$ define

$$T_{n,j}(p^2) = \Gamma_n \mathrm{diag}(I^{(n-j)}, pI^{(j)}, p^2 I^{(n-j)}, pI^{(j)})\Gamma_n \subset \Delta_n(p^2) . \tag{4}$$

Due to (1)

$$\Delta_n(p^2) = \bigcup_{j=0}^{n} T_{n,j}(p^2) \tag{5}$$

holds. Recall the definition of the p-rank $r_p(M)$ of an integral matrix M from section V.7 . Again we get the simple but effective criterion:

(5.2) Lemma. *Given* $M \in \Delta_n(p^2)$ *one has*

$$M \in T_{n,j}(p^2) \quad \textit{if and only if} \quad r_p(M) = n-j .$$

Using (5.1) we obtain the explicit formula

(5.3) Lemma. $T_n(p)^2 = T_{n,0}(p^2) + \sum_{j=1}^{n} (\mathrm{ind}\ T_j(p))\ T_{n,j}(p^2)$.

Proof. According to (5) the definition of the product leads to

$$T_n(p)^2 = \sum_{j=0}^{n} t_j\ T_{n,j}(p^2) ,$$

where t_j equals the number of representatives $M:\Gamma_n \backslash T_n(p)$ such that $M_j M^{-1}$ is integral, where we choose the representative

$$M_j = \mathrm{diag}(p^2 I^{(n-j)}, pI^{(j)}, I^{(n-j)}, pI^{(j)}) .$$

of $T_{n,j}(p^2)$ in place of the canonical one in (4) . Using (3.3) , (1.1) , and the description above we conclude that this holds if and only if the representative M has the form

$$M = \begin{pmatrix} pI & 0 \\ 0 & I \end{pmatrix} \times M_1 ,$$

where $I = I^{(n-j)}$ and M_1 runs through a set of representatives of $\Gamma_j \backslash T_j(p)$, because pM_1^{-1} is integral again.

□

(5.4) Exercises. a) Let $0 \le j \le n$. Which elementary divisor matrices are contained in $\mathcal{D}(T_{n,j}(p^2))$?

b) Derive the analogue for $\mathcal{H}^{\mathfrak{o}}_{n,p}$, whenever $\mathfrak{o}$ has the class number 1.

c) Demonstrate the analogue for $\mathcal{H}^{*}_{n,p}$, $n \ge 1$.

6. The Primary Component

In this section we determine the structure of the primary components. Fix a prime number p. As for the unimodular group we conclude from 1(1) that the Hecke algebra $\mathcal{H}_{n,p}$ is a $\mathbb{Z}$-graded algebra. One has

$$\mathcal{H}_{n,p} = \bigoplus_{\ell=0}^{\infty} \mathcal{H}^{(\ell)}_{n,p} ,$$

where

$$\mathcal{H}^{(\ell)}_{n,p} = \left\{ \sum_{M:\Gamma_n\backslash\Delta_n(p^\ell)/\Gamma_n} t(\Gamma_n M\Gamma_n)\,\Gamma_n M\Gamma_n \mid t(\Gamma_n M\Gamma_n) \in \mathbb{Z} \right\}.$$

In particular we get

(1) $$\mathcal{H}^{(0)}_{n,p} = \mathbb{Z}\,\Gamma_n ,$$

(2) $$\mathcal{H}^{(1)}_{n,p} = \mathbb{Z}\,T_n(p) ,$$

(3) $$\mathcal{H}^{(k)}_{n,p}\cdot\mathcal{H}^{(\ell)}_{n,p} \subset \mathcal{H}^{(k+\ell)}_{n,p} .$$

A result on the generators of $\mathcal{H}_{n,p}$ is derived in

(6.1) Proposition. *The $\mathbb{Z}$-algebra $\mathcal{H}_{n,p}$ is generated by*

$$T_n(p),\ T_{n,1}(p^2),\ T_{n,2}(p^2),\dots,T_{n,n}(p^2)$$

as a unitary algebra.

Proof. Let $\mathcal{A}$ denote the unitary subalgebra of $\mathcal{H}_{n,p}$, which is generated by the double cosets $T_n(p), T_{n,1}(p^2),\dots,T_{n,n}(p^2)$. We show by induction on ℓ that

(4) $$\mathcal{H}^{(\ell)}_{n,p} \subset \mathcal{A}$$

holds. The cases $\ell = 0,1$ are obvious due to (1), (2), and $\ell = 2$ follows from (5.3) and the identity

$$\mathcal{H}^{(2)}_{n,p} = \mathbb{Z}\,T_{n,0}(p^2) + \mathbb{Z}\,T_{n,1}(p^2) + \dots + \mathbb{Z}\,T_{n,n}(p^2) .$$

Given $\ell \ge 3$ we show successively that the submodule

$$\mathcal{H}^{(\ell,r)}_{n,p} = \left\{ \sum_{\substack{M:\Gamma_n\backslash\Delta_n(p^\ell)/\Gamma_n \\ r_p(M)=r}} t(\Gamma_n M\Gamma_n)\,\Gamma_n M\Gamma_n \mid t(\Gamma_n M\Gamma_n) \in \mathbb{Z} \right\},$$

$0 \le r \le n$, is contained in $\mathcal{A}$. Let a double coset $\Gamma_n M \Gamma_n \in \mathcal{H}_{n,p}^{(\ell,r)}$ be given and choose the representative M in the form

(5) $$M = \operatorname{diag}(p^{k_1},\dots,p^{k_n},p^{\ell-k_1},\dots,p^{\ell-k_n}),\ 0 = k_1 = \dots = k_r < k_{r+1} \le \dots \le k_n \le \ell/2 .$$

$r = 0$ is equivalent with $k_1 \ge 1$. In this case

$$T := \Gamma_n(\tfrac{1}{p}M)\Gamma_n \in \mathcal{A}$$

holds by induction hypothesis. In view of $T_{n,n}(p^2) = \Gamma_n(pI)\Gamma_n$ now (4.1) leads to

$$\Gamma_n M \Gamma_n = T \cdot T_{n,n}(p^2) \in \mathcal{A} .$$

If $0 < r \le n$ consider

$$T := \Gamma_n \operatorname{diag}(1,\dots,1,p^{k_{r+1}-1},\dots,p^{k_n-1},p^{\ell-2},\dots,p^{\ell-2},p^{\ell-k_{r+1}-1},\dots,p^{\ell-k_n-1})\Gamma_n .$$

We have $T \in \mathcal{A}$ and calculate

$$T \cdot T_{n,n-r}(p^2) = \sum_{N:\Gamma_n\backslash\Delta_n(p^\ell)/\Gamma_n} t(\Gamma_n N \Gamma_n)\, \Gamma_n N \Gamma_n \in \mathcal{H}_{n,p}^{(\ell)} .$$

At first $t(\Gamma_n N \Gamma_n) \neq 0$ implies $r_p(N) \le r$ due to (5.2). Consider $\Gamma_n N \Gamma_n \in \mathcal{H}_{n,p}^{(\ell,r)}$ with a representative

$$N = \operatorname{diag}(p^\ell,\dots,p^\ell,p^{\ell-j_{r+1}},\dots,p^{\ell-j_n},1,\dots,1,p^{j_{r+1}},\dots,p^{j_n}),$$
where $1 \le j_{r+1} \le \dots \le j_n \le \ell/2$,

such that JNJ^{-1} has the elementary divisor normal form (2.2). We take representatives K of $\Gamma_n\backslash T_{n,n-r}(p^2)$ in the form (3.3). Then $t(\Gamma_n N \Gamma_n)$ equals the number of K's such that

$$NK^{-1} \in T .$$

Choosing for D representatives in the form V(5.1) the form of N immediately yields

$$K = \begin{pmatrix} p^2 I & 0 \\ 0 & I \end{pmatrix} \times K_1 ,\quad I = I^{(r)} ,$$

for some $K_1 \in \Delta_{n-r}(p^2)$. But then (5.2) implies $K_1 = pI$. Thus there is only one representative $K = \operatorname{diag}(p^2 I^{(r)}, pI^{(n-r)}, I^{(r)}, pI^{(n-r)})$ for which NK^{-1} is integral again. But $JNK^{-1}J^{-1}$ has the elementary divisor normal form (2.2). Thus $N = JMJ^{-1}$ follows and we obtain

$$T \cdot T_{n,n-r}(p^2) = T_1 + \Gamma_n M \Gamma_n ,$$

where

$$T_1 \in \bigoplus_{s=0}^{r-1} \mathcal{H}_{n,p}^{(\ell,s)} .$$

Because of $\mathcal{T}_{n,0}(p^2) \in \mathcal{A}$ the induction hypothesis yields

$$\mathcal{A} = \mathcal{H}_{n,p} . \qquad \square$$

The structure of the primary component is now described in

(6.2) Theorem. *$\mathcal{H}_{n,p}$ equals the polynomial ring over $\mathbb{Z}$ in $n+1$ elements*

$$\mathcal{T}_n(p),\ \mathcal{T}_{n,1}(p^2),\ldots,\mathcal{T}_{n,n}(p^2)\ ,$$

which are algebraically independent. In particular $\mathcal{H}_n$ does not contain any zero-divisors.

Proof. (6.1) yields

$$\mathcal{H}_{n,p} = \mathbb{Z}\,[\,\mathcal{T}_n(p),\ \mathcal{T}_{n,1}(p^2),\ldots,\mathcal{T}_{n,n}(p^2)\,]\ .$$

In view of (1) elementary combinatorics lead to

$$\dim \mathcal{H}_{n,p}^{(\ell)} = \#\left\{ (k_1,\ldots,k_n)' \in \mathbb{Z}^n \mid 0 \le k_1 \le \ldots. \le k_n \le \ell/2 \right\} = \binom{n+[\ell/2]}{n}\ .$$

Due to (3) a monomial

$$\mathcal{T}_n^{\,m_0} \cdot \mathcal{T}_{n,1}^{\,m_1} \cdot \ldots \cdot \mathcal{T}_{n,n}^{\,m_n}$$

belongs to $\mathcal{H}_{n,p}^{(\ell)}$ if and only if

$$m_0 + 2m_1 + 2m_2 + \ldots + 2m_n = \ell\ .$$

Again elementary combinatorics show that the number of these monomials also equals $\binom{n+[\ell/2]}{n}$. Since the monomials span $\mathcal{H}_{n,p}^{(\ell)}$, they are linearly independent. Hence $\mathcal{T}_n(p), \mathcal{T}_{n,1}(p^2),\ldots,\mathcal{T}_{n,n}(p^2)$ are algebraically independent. $\square$

It is easy to derive the structure of the Hecke algebra $\bar{\mathcal{H}}_n = \mathcal{H}(\Gamma_n,\bar{\Delta}_n)$ from (6.2). One obtains

$$\bar{\mathcal{H}}_n = \bigotimes_{p \text{ prime}} \bar{\mathcal{H}}_{n,p}\ ,$$

where

$$\bar{\mathcal{H}}_{n,p} = \mathbb{Z}[\,\mathcal{T}_n(p),\ \mathcal{T}_{n,1}(p^2),\ldots,\mathcal{T}_{n,n}(p^2),\ \mathcal{T}_{n,n}(p^2)^{-1}\,]\ .$$

E. Freitag [18] and A.N. Andrianov [4] prove (6.2) by induction on n. Therefore a homomorphism is used, which will be described in the sequel. Given $n > 1$ we conclude from (3.5) and (1.2) that each right coset $\Gamma_n M$, $M \in \Delta_n(q)$, possesses a representative $M \in \Delta_{n,n-1}$ of the form

$$M = \begin{pmatrix} A & B \\ C & D \end{pmatrix},\ A = \begin{pmatrix} A_1 & 0 \\ a' & \alpha \end{pmatrix},\ B = \begin{pmatrix} B_1 & * \\ * & * \end{pmatrix},\ C = \begin{pmatrix} C_1 & 0 \\ 0 & 0 \end{pmatrix},\ D = \begin{pmatrix} D_1 & d \\ 0 & \delta \end{pmatrix},$$

$$M_1 = \begin{pmatrix} A_1 & B_1 \\ C_1 & D_1 \end{pmatrix} \in \Delta_{n-1}(q) .$$

Then we define

$$\psi(\Gamma_n M) := \begin{cases} \Gamma_{n-1} M_1 & \text{if } \delta = \pm 1 \\ 0 & \text{otherwise} \end{cases} \in \mathcal{R}_{n-1} = \mathcal{R}(\Gamma_{n-1}, \Delta_{n-1}) .$$

If $\delta = \pm 1$ one can choose the representative M in the form

(6) $$M = M_1 \times \begin{pmatrix} q & 0 \\ 0 & 1 \end{pmatrix}$$

due to (3.4) . Clearly the definition does not depend on the choice of the representative M in $\Delta_n(q) \cap \Delta_{n,n-1}$. The mapping ψ can be linearly extended to $\mathcal{R}_n$. Multiplication by matrices $K_1 \times I^{(2)}$, $K_1 \in \Gamma_{n-1}$, from the right hand side yields

$$\psi(\mathcal{R}_n^{\Gamma_n}) \subset \mathcal{R}_{n-1}^{\Gamma_{n-1}} , \quad \text{hence } \psi(\mathcal{H}_n) \subset \mathcal{H}_{n-1}$$

due to I(3.9) . The definition of the action and (6) immediately lead to

$$\psi(T * U) = \psi(T) * \psi(U) \quad \text{for } T \in \mathcal{H}_n , \; U \in \mathcal{R}_n .$$

Thus ψ is a homomorphism of the Hecke algebras.

(6.3) Lemma. *If* $n > 1$, *the homomorphism of the algebras*

$$\psi : \mathcal{H}_n \to \mathcal{H}_{n-1}$$

is surjective. Given a prime number p *one has*

(7) $$\psi(T_n(p)) = T_{n-1}(p) , \quad \psi(T_{n,j}(p^2)) = T_{n-1,j}(p^2) \quad \textit{for } 0 \le j < n .$$

Moreover the kernel of $\psi|_{\mathcal{H}_{n,p}}$ *is the principal ideal generated by* $T_{n,n}(p^2)$.

Proof. (7) follows from (3.4) and (5.2) . Hence ψ is surjective according to (6.2) and (4.2) . Clearly $T_{n,n}(p)$ belongs to kernel ψ . If we combine this with (7) and (6.2) , we get

$$\text{kernel } \psi|_{\mathcal{H}_{n,p}} = T_{n,n}(p) \cdot \mathcal{H}_{n,p} . \qquad \square$$

(6.4) Remark. There also exists a description of the Hecke algebra in terms of polynomials just as in section V.10 in the case of the unimodular group. Recall the definition of the P-polynomial from V.10 . Given an additional independent indeterminate X_0 and $M \in \Delta_n(p^\ell)$, we define

$$Q(\Gamma_n M \Gamma_n) := \sum_{\begin{pmatrix} A & B \\ 0 & D \end{pmatrix} : \Gamma_n \backslash \Gamma_n M \Gamma_n} |\det A|^{n+1} X_0^{-\ell} P(\mathcal{U}_n A) .$$

It can be shown that Q gives an isomorphism between $\bar{\mathcal{H}}_n \otimes_{\mathbb{Z}} \mathbb{Q}$ and a particular subalgebra of the polynomial algebra $\mathbb{Q}[X_0^{\pm 1},\ldots,X_n^{\pm 1}]$, which is characterized by the invariance under the Weyl group of the symplectic group. This is pointed out in [4] , chapter 3 , and in [18] , chapter 4 .

(6.5) Exercises. a) Given a prime p and $0 \le j \le n$ define the double coset

$$\widetilde{T}_{n,j}(p) := \Gamma_n \mathrm{diag}\left(\tfrac{1}{p}I^{(n-j)}, I^{(j)}, pI^{(n-j)}, I^{(j)}\right)\Gamma_n = T_{n,j}(p^2)\cdot T_{n,n}(p^2)^{-1} .$$

Show that the Hecke algebra $\mathcal{H}(\Gamma_n, Sp(n;\mathbb{Q}))$ over $\mathbb{Z}$ equals the polynomial ring over $\mathbb{Z}$ in the elements

$$\widetilde{T}_{n,j}(p) \ , \ 0 \le j < n \ , \text{ where } p \text{ runs through all primes,}$$

which are algebraically independent.
b) Express the double cosets in $\Delta_2(p^3)$ and $\Delta_3(p^3)$ explicitly as polynomials in the generating double cosets in (6.1) .
c) Consider the Q-polynomials in (6.4) . Show that

$$Q(T_n(p)) = p^{n(n+1)/2}\, X_0^{-1}\prod_{j=1}^{n}(X_j+1) ,$$

$$Q(T_{n,n}(p^2)) = p^{n(n+1)/2}\, X_0^{-2}X_1X_2\ldots X_n .$$

d) Let $\Gamma_{n,0}$ resp. $\Delta_{n,0}$ denote the subgroup of Γ_n resp. the submonoid of Δ_n , where the C-block of each matrix is equal to 0 . Show that $(\Gamma_{n,0},\bar{\Delta}_{n,0})$ is a Hecke pair and that the Hecke algebra $\mathcal{H}(\Gamma_{n,0},\Delta_{n,0})$ fails to be commutative and contains zero-divisors. Show that there exists a canonical embedding $\nu : \mathcal{H}(\Gamma_n,\Delta_n) \longrightarrow \mathcal{H}(\Gamma_{n,0},\Delta_{n,0})$.
e) Derive the structure theorem for the Hecke algebra $\mathcal{H}^{o}_{n,p}$, whenever o has the class number 1 .
f) Determine the analogue of (6.2) for $\mathcal{H}^{*}_{n,p}$, $n \ge 1$.

7. The Commutation Relation with the Siegel Φ-Operator

In this section we investigate the homomorphism of the Hecke algebras, which describes the commutation relation with the Siegel Φ-operator (cf. [4] , [18]) . We will show that this homomorphism is surjective by computing the images of the generating double cosets in (6.2) .

Therefore we have to consider the Hecke algebra over $\mathbb{Q}$, i.e.,

$$\widetilde{\mathcal{H}}_n = \mathcal{H}_n \otimes_{\mathbb{Z}} \mathbb{Q} \ , \quad \widetilde{\mathcal{R}}_n = \mathcal{R}(\Gamma_n,\Delta_n)\otimes_{\mathbb{Z}}\mathbb{Q} .$$

Given $n > 1$ and a right coset $\Gamma_n M \subset \Delta_n(q)$ we choose a representative M in $\Delta_{n,n-1}$, i.e.

$$M = \begin{pmatrix} A & B \\ C & D\end{pmatrix} , \ A = \begin{pmatrix} A_1 & 0 \\ a' & \alpha\end{pmatrix} , \ B = \begin{pmatrix} B_1 & * \\ * & *\end{pmatrix} , \ C = \begin{pmatrix} C_1 & 0 \\ 0 & 0\end{pmatrix} , \ D = \begin{pmatrix} D_1 & d \\ 0 & \delta\end{pmatrix} ,$$

$$M_1 = \begin{pmatrix} A_1 & B_1 \\ C_1 & D_1 \end{pmatrix} \in \Delta_{n-1}(q) .$$

Given an even integer r define

$$\Phi_r(\Gamma_n M) = \delta^{-r}\, \Gamma_{n-1} M_1 \in \tilde{\mathcal{R}}_{n-1} .$$

Clearly the definition does not depend on the choice of the representative M in $\Delta_{n,n-1}$ due to (1.2), because r is even and $M \in \Gamma_{n,n-1}$ yields $\delta = \pm 1$. The mapping Φ_r is linearly extended to $\tilde{\mathcal{R}}_n$. Multiplication by matrices $K_1 \times I^{(2)}$, $K_1 \in \Gamma_{n-1}$, from the right hand side yields

$$\Phi_r\left(\tilde{\mathcal{R}}_n^{\Gamma_n}\right) \subset \tilde{\mathcal{R}}_{n-1}^{\Gamma_{n-1}} , \text{ hence } \Phi_r(\tilde{\mathcal{H}}_n) \subset \tilde{\mathcal{H}}_{n-1}$$

due to I(3.9). The definition of the action leads to

$$\Phi_r(T * U) = \Phi(T) * \Phi(U) \quad \text{for } T \in \tilde{\mathcal{H}}_n , U \in \tilde{\mathcal{R}}_n .$$

Thus Φ_r is a homomorphism of the Hecke algebras. We calculate the images of the generators explicitly.

(7.1) Theorem. *Let* p *be a prime number,* r *an even integer, and* $n > 1$. *Then one has*

(1) $$\Phi_r(T_n(p)) = (p^{n-r}+1)\, T_{n-1}(p) ,$$

(2) $$\Phi_r(T_{n,j}(p^2)) = \sum_{\substack{i=j-1,j,j+1 \\ 0 \le i \le n-1}} t_i\, T_{n-1,i}(p^2)$$

for $0 \le j \le n$, *where*

$$t_{j-1} = p^{n-j-r} , \quad t_j = 1+(p-1)p^{n-j-1-r} + p^{2n-2r} , \quad t_{j+1} = (p^{2j+2}-1)p^{n-j-1-r} .$$

Proof. A set of representatives of the right coset $\Gamma_n M \subset \Delta_n(p)$, where $\Phi_r(\Gamma_n M)$ is a multiple of $\Gamma_{n-1}\begin{pmatrix} I & 0 \\ 0 & pI \end{pmatrix}$, $I = I^{(n-1)}$, is given by

$$M = \begin{pmatrix} A & B \\ 0 & D \end{pmatrix} , \quad A = pD'^{-1} ,$$

$$D = \begin{pmatrix} pI & 0 \\ 0 & 1 \end{pmatrix} , \quad B = 0 \quad \text{and} \quad D = \begin{pmatrix} pI & 0 \\ 0 & p \end{pmatrix} , \quad B = \begin{pmatrix} 0 & b \\ b' & \beta \end{pmatrix} , \begin{pmatrix} b \\ \beta \end{pmatrix} \in \mathbb{Z}^n \bmod p$$

according to (3.4) and 5(2). This yields (1).

Let $0 \le k \le n-1$, $\ell = n-1-k$, and

$$N = \operatorname{diag}\left(I^{(k)}, pI^{(\ell)}, p^2 I^{(k)}, pI^{(\ell)}\right) \in \Delta_{n-1}(p^2) .$$

Now we determine all the right cosets $\Gamma_n M \subset \Delta_n(p^2)$ such that $\Phi_r(\Gamma_n M)$ is a multiple of $\Gamma_{n-1} N$. The form of the last columm of D follows from $p^2 D'^{-1} \in Mat(n;\mathbb{Z})$, the form of B from $D'B = B'D$ and $B_1 = 0$. An explicit set of representatives is given by

$$M = \begin{pmatrix} A & B \\ 0 & D \end{pmatrix} , \quad A = p^2 D'^{-1} ,$$

where

(3) $$D = \mathrm{diag}\left(p^2 I, pI, 1\right) \ , \quad B = 0 \ ,$$

(4) $$D = \begin{pmatrix} p^2 I & 0 & 0 \\ 0 & pI & d \\ 0 & 0 & p \end{pmatrix} \ , \quad B = \begin{pmatrix} 0 & 0 & b_1 \\ 0 & 0 & b_2 \\ pb_1' & b_2' & \beta \end{pmatrix} \ , \quad \begin{array}{l} b_1 \in \mathbb{Z}^k \bmod p \ , \quad \beta \in \mathbb{Z} \bmod p \ , \\ b_2, d \in \mathbb{Z}^\ell \bmod p \ , \end{array}$$

(5) $$D = \begin{pmatrix} p^2 I & 0 & 0 \\ 0 & pI & pd \\ 0 & 0 & p^2 \end{pmatrix} \ , \quad B = \begin{pmatrix} 0 & 0 & b_1 \\ 0 & 0 & pb_2 \\ b_1' & b_2' & \beta \end{pmatrix} \ , \quad \begin{array}{l} b_1 \in \mathbb{Z}^k \bmod p^2 \ , \quad \beta \in \mathbb{Z} \bmod p^2 \ , \\ b_2, d \in \mathbb{Z}^\ell \bmod p \ . \end{array}$$

Using (5.2) we verify that the matrix (3) as well as the p^{2n} matrices (5) belong to $T_{n,\ell}(p^2)$. In (4) there are p^k matrices with $(d', b_2', \beta) \equiv 0 \bmod p$, which belong to $T_{n,\ell+1}(p^2)$. All the $p^k(p-1)$ matrices with $(d', b_2') \equiv 0 \bmod p$ and $\beta \not\equiv 0 \bmod p$ are contained in $T_{n,\ell}(p^2)$. Finally all the $(p^{2\ell}-1)p^{k+1}$ matrices with $(d', b_2') \not\equiv 0 \bmod p$ belong to $T_{n,\ell-1}(p^2)$.

Using (3) , (4) , (5) we have described the inverse image of $\Gamma_{n-1}N$ under the map $\Phi_r : \tilde{\mathcal{R}} \rightarrow \tilde{\mathcal{R}}_{n-1}$. Now we determine (2) by calculating the coefficient of the particular right coset $\Gamma_{n-1}N$ in $\Gamma_{n-1}N\Gamma_{n-1}$. □

A combination of (7.1) , (6.2) , and (4.2) yields the announced result on the surjectivity of the homomorphism.

(7.2) Corollary. *Given* $n > 1$ *and an even integer* r *the homomorphism*

$$\Phi_r \ : \ \tilde{\mathcal{H}}_n \rightarrow \tilde{\mathcal{H}}_{n-1}$$

is surjective.

The theorem also allows us to calculate the number of right cosets in each generating double coset, since the theorem applied to the case $r = 0$ gives a recursion formula for these numbers, which can be solved explicitly.

(7.3) Corollary. *One has*

$$\mathrm{ind}\ T_{n,n}(p^2) = 1$$

and for $0 \le j < n$

$$\mathrm{ind}\ T_{n,j}(p^2) = p^{(n-j)(n+1-j)/2} \prod_{i=1}^{n-j} \frac{p^{2(n+1-i)}-1}{p^i-1} \ .$$

Proof. $\mathrm{ind}\ T_{n,n}(p^2) = 1$ follows from $T_{n,n}(p^2) = \Gamma_n(pI)\Gamma_n$. If we set $r = 0$ in (7.1) , we get

$$\mathrm{ind}\ T_{n,j}(p^2) = \sum_{\substack{i=j-1,j,j+1 \\ 0 \le i \le n-1}} t_i \ \mathrm{ind}\ T_{n-1,i}(p^2) \ ,$$

where

$$t_{j-1} = p^{n-j} \quad , \quad t_j = p^{2n} + (p-1)p^{n-j-1} + 1 \quad , \quad t_{j+1} = \left(p^{2j+2} - 1\right) p^{n-j-1} \; .$$

A simple induction shows that the values above satisfy this recursion formula. The case $n = 1$ follows from IV(3.3) . □

(7.4) Exercises. a) Suppose that r is divisible by the number of units in $\mathfrak{o}$. Then derive the analogue of (7.1) for $\mathcal{H}^{\mathfrak{o}}_{n,p}$, if $\mathfrak{o}$ has the class number 1 .
b) Determine the analogue of (7.1) for $\mathcal{H}^{*}_{n,p}$.

8. A Rationality Theorem for $n = 2$

In this section we derive the analogue of IV(4.3) for $n = 2$. The result is due to G. Shimura [57] (cf. also A.N. Andrianov [4]). For arbitrary n a corresponding result over local fields was proved by A.N. Andrianov [2] , [3] .

Given $q \in \mathbb{N}$ recall the definition of

$$T_n(q) := \sum_{M:\Gamma_n\backslash\Delta_n(q)/\Gamma_n} \Gamma_n M \Gamma_n \, .$$

Let p be a prime, $\ell \geq 0$, and $0 \leq r \leq n$. Then set

$$T_n^{(r)}(p^\ell) := \sum_{\substack{M:\Gamma_n\backslash\Delta_n(p^\ell)/\Gamma_n \\ r_p(M)=r}} \Gamma_n M \Gamma_n \, .$$

Note that

$$T_n(1) = \Gamma_n \, , \quad T_n(p) = T_n^{(n)}(p) \, , \quad T_{n,j}(p^2) = T_n^{(n-j)}(p^2) \quad \text{for } 0 \leq j \leq n \, ,$$

(1) $$T_n(p^\ell) = \sum_{r=0}^{n} T_n^{(r)}(p^\ell) \quad \text{for } \ell \geq 1 \, .$$

Furthermore (4.1) yields

(2) $$T_n^{(0)}(p^{\ell+2}) = T_{n,n}(p^2) \cdot T_n(p^\ell) \quad \text{for } \ell \geq 0 \, .$$

With the formal extension $\operatorname{ind} T_0(p) := 1$ a slight modification of the proof of (5.3) yields

(8.1) Lemma. *Given a prime* p *and* $\ell \in \mathbb{N}$

$$T_n(p) \cdot T_n(p^\ell) = \sum_{r=0}^{n} \left(\operatorname{ind} T_{n-r}(p)\right) T_n^{(r)}(p^{\ell+1}) \, .$$

holds. In particular one has

(3) $$T_2(p) \cdot T_2(p^\ell) = T_2(p^{\ell+1}) + p\, T_2^{(1)}(p^{\ell+1}) + \left(p^3 + p^2 + p\right) T_{2,2}(p^2) \cdot T_2(p^{\ell-1}) \, .$$

Proof. The definition of the product yields

$$T_n(p^\ell)\cdot T_n(p) = \sum_{M:\Gamma_n\backslash\Delta_n(p^{\ell+2})/\Gamma_n} t(\Gamma_n M\Gamma_n)\,\Gamma_n M\Gamma_n ,$$

where $t(\Gamma_n M\Gamma_n)$ equals the number of representatives $L:\Gamma_n\backslash T_n(p)$ such that ML^{-1} is integral. If $r_p(M) = r$ choose the representative M in the form

$$M = M_1\times\begin{pmatrix} p^{\ell+1}I & 0 \\ 0 & I\end{pmatrix} , \quad I = I^{(r)} , \quad M_1 \in \Delta_{n-r}(p^{\ell+1}) , \quad r_p(M_1) = 0 .$$

Using (3.4) and the results of section 5 we conclude that L has the form

$$L = L_1\times\begin{pmatrix} pI & 0 \\ 0 & I\end{pmatrix}, \quad I = I^{(r)} ,$$

where L_1 runs through a set of representatives of $\Gamma_{n-r}\backslash T_{n-r}(p)$. The use of (1) , (2) , and (5.1) yields (3) . □

Now we have a closer look at $n = 2$. In this case we compute the products of $T_2(p^\ell)$ with all the generating double cosets explicitly.

(8.2) Lemma. *Given a prime* p *and* $\ell \geq 2$ *the following hold*

$$T_{2,1}(p^2)\cdot T_2(p) = T_2^{(1)}(p^3) + (p^2+p)\,T_{2,2}(p^2)\cdot T_2(p) ,$$

$$T_{2,1}(p^2)\cdot T_2(p^\ell) = T_2^{(1)}(p^{\ell+2}) + (p^2+p)\,T_{2,2}(p^2)\cdot T_2(p^\ell)$$
$$+\, p^3\,T_{2,2}(p^2)\cdot T_2^{(1)}(p^\ell) + (p^4+p^3)\,T_{2,2}(p^2)^2\cdot T_2(p^{\ell-2}) .$$

Proof. The same arguments as for (8.1) and (4.1) lead to

$$T_{2,1}(p^2)\cdot T_2(p^\ell) = T_2^{(1)}(p^{\ell+2}) + \sum_{\substack{N:\Gamma_2\backslash\Delta_2(p^{\ell+2})/\Gamma_2 \\ r_p(N)=0}} \bar{t}(\Gamma_2 N\Gamma_2)\,\Gamma_2 N\Gamma_2$$
$$= T_2^{(1)}(p^{\ell+2}) + T_{2,2}(p^2) \sum_{M:\Gamma_2\backslash\Delta_2(p^\ell)/\Gamma_2} t(\Gamma_2 M\Gamma_2)\,\Gamma_2 M\Gamma_2$$

for $\ell \geq 1$. Let $M \in \Delta_2(p^\ell)$ with $r_p(M) = 2$. Hence we may choose

$$M = \operatorname{diag}(p^\ell, p^\ell, 1, 1)$$

as a representative. Then a set of representatives L of $\Gamma_2\backslash T_{2,1}(p^2)$ such that pML^{-1} is integral is given by

(4) $$\begin{pmatrix} A & 0 \\ 0 & D\end{pmatrix} , \quad D : \mathcal{U}_2\backslash T_{2,1}(p) , \quad A = p^2 D'^{-1}$$

and

(5) $$\begin{pmatrix} pI & S \\ 0 & pI\end{pmatrix} , \quad S \in Sym(2;\mathbb{Z}) \bmod p , \quad r_p(S) = 1$$

according to (3.4) and (5.2). Thus

$$t(\Gamma_2 M \Gamma_2) = (p+1) + (p^2-1) = p^2 + p$$

follows. Now let $r_p(M) = 1$ and therefore $\ell \geq 2$. Hence we may choose

$$M = \operatorname{diag}(p^{\ell-m}, p^{\ell}, p^m, 1), \quad 0 < m \leq \ell/2 .$$

Then a set of representatives L of $\Gamma_2 \backslash T_{2,1}(p^\ell)$ such that pML^{-1} is integral is given by the matrices (4), (5), and additionally by

$$\begin{pmatrix} 1 & 0 & b_1 & b_2 \\ 0 & p & pb_2 & 0 \\ 0 & 0 & p^2 & 0 \\ 0 & 0 & 0 & p \end{pmatrix}, \qquad \begin{array}{l} b_1 = 0,1,\ldots,p^2-1 , \\ \\ b_2 = 0,1,\ldots,p-1 . \end{array}$$

Thus

$$t(\Gamma_2 M \Gamma_2) = p^3 + p^2 + p$$

follows. If $r_p(M) = 0$ and therefore $\ell \geq 2$, one easily checks

$$t(\Gamma_2 M \Gamma_2) = \operatorname{ind} T_{2,1}(p^2) = p^4 + p^3 + p^2 + p$$

due to (7.3). This yields the assertion. □

Now we combine all the results above in order to obtain the rationality of a formal power series over the Hecke algebra $\mathcal{H}_2$. A straightforward calcution leads to

(8.3) Theorem. *Rationality Theorem*

Let p *be a prime and* X *an indeterminate. Then*

$$\sum_{\ell=0}^{\infty} T_2(p^\ell)\, X^\ell = \left(1 - p^2\, T_{2,2}(p^2)\, X^2\right) \cdot R(X)^{-1}$$

holds formally in X, *where*

$$R(X) = 1 - T_2(p)\,X + [(p^3+p)\,T_{2,2}(p^2) + p\,T_{2,1}(p^2)]\,X^2 - p^3\,T_{2,2}(p^2)\cdot T_2(p)\,X^3 + p^6\,T_{2,2}(p^2)^2\,X^4 .$$

Proof. Compute

$$R(X) \cdot \sum_{\ell=0}^{\infty} T_2(p^\ell)\, X^\ell .$$

Use (8.1), (8.2), (1), and (2). □

(8.4) Remark. a) It follows from (8.1) that the coefficient of X^2 in $R(X)$ equals

$$(p^3+p)\,T_{2,2}(p^2) + p\,T_{2,1}(p^2) = T_2(p)^2 - T_2(p^2) - p^2\,T_{2,2}(p^2) .$$

b) In the case $n = 1$ the Rationality Theorem was already proved in IV(4.3). Hence

$$\sum_{\ell=0}^{\infty} T_1(p^\ell) X^\ell = \left(1 - T_1(p) X + p\, T_{1,1}(p^2) X^2\right)^{-1}$$

holds as a formal power series in X. This result also follows directly from (8.3) by application of the mapping ψ in (6.3).

(8.5) Exercises. a) Consider the embedding $\nu : \mathcal{H}(\Gamma_2,\Delta_2) \to \mathcal{H}(\Gamma_{2,0},\Delta_{2,0})$ in (6.5). Show that the image of the polynomial $R(X)$ in (8.3) under the embedding ν decomposes in $\mathcal{H}(\Gamma_{2,0},\Delta_{2,0})[X]$ (cf. [4], 3.5.14).
b) Derive the analogue of (8.3) in $\mathcal{H}^{\mathfrak{o}}_{2,p}$, where $\mathfrak{o}$ has the class number 1, whenever p is inert or if p is ramified in $\mathfrak{o}$.
c) Determine the analogue of (8.3) for $\mathcal{H}^{*}_{2,p}$.

Remarks

The structure of the Hecke algebra associated with the Siegel modular group was determined by G. Shimura [57]. The results of this chapter are also contained in the books of A.N. Andrianov [4] and E. Freitag [18]. They include the description in terms of the Q-polynomials (cf. (6.4)), which is due to I. Satake [53]. The result (2.1) on the generators of the Siegel modular group is due to L. Kronecker [41]. It is often called Witt's Theorem [65].

The commutation relation between Hecke operators and the Siegel Φ-operator was derived by N.A. Zharkovskaya [66] on the level of the Q-polynomials. The explict images of the generators were computed in [35].

The Rationality Theorem in section 8 is due to G. Shimura [57]. A.N. Andrianov [2], [3] demonstrated a Rationality Theorem over local fields. Recently S. Böcherer [8] derived a Rationality Theorem for the subalgebra $\mathcal{H}(\Gamma_n,Sp(n;\mathbb{Q}))$ in terms of the Q-polynomials.

References

[1] Aitken, A.C.: Determinants and Matrices, 7th edition. Oliver and Boyd, Edinburgh-London 1951.

[2] Andrianov, A.N.: Rationality theorems for Hecke series and zeta-functions of the groups GL_n and Sp_n over local fields. Math. USSR Izv. **3** (1969), 439-476.

[3] Andrianov, A.N.: Spherical functions for GL_n over local fields and summation of Hecke series. Math. USSR Sb. **12** (1970), 429-452.

[4] Andrianov, A.N.: Quadratic Forms and Hecke Operators. Grundlehren Math. Wiss. **286**, Springer-Verlag, Berlin-Heidelberg-New York 1987.

[5] Apostol, T.M.: Introduction to Analytic Number Theory. Springer-Verlag, Berlin-Heidelberg-New York 1976.

[6] Apostol, T.M.: Modular Functions and Dirichlet Series in Number Theory. Springer-Verlag, Berlin-Heidelberg-New York 1976.

[7] Artin, E.: Geometric Algebra. Interscience Publishers, New York 1957.

[8] Böcherer, S.: Ein Rationalitätssatz für formale Heckereihen zur Siegelschen Modulgruppe. Abh. Math. Sem. Univ. Hamburg **56** (1986), 35-47.

[9] Bourbaki, N.: Groupès et Algèbres de Lie, Chap. IV. Hermann, Paris 1968.

[10] Braun, H.: Hermitian Modular Functions I, II, III. Ann. Math., II. Ser. **50** (1949), 827-855; Ann. Math., II. Ser. **51** (1950), 92-104; Ann. Math., II. Ser. **53** (1951), 143-160.

[11] Braun, H., and Koecher, M.: Jordan-Algebren. Grundlehren Math. Wiss. **128**, Springer-Verlag, Berlin-Heidelberg-New York 1966.

[12] Bump, D.: Automorphic Forms on $GL(3,\mathbb{R})$. Lect. Notes Math. **1083**, Springer-Verlag, Berlin-Heidelberg-New York 1984.

[13] Cayley, A.: On the determination of the value of a certain determinant.Quarterly Math. J. **2** (1858), 163-166; Collected Math. Papers III, 120-123.

[14] Curtis, C.W.: Representations of finite groups of Lie type. Bull. Am. Math. Soc., New Ser. **1** (1979), 721-757.

[15] Curtis, C.W., Iwahori, N., and Kilmoyer, R.: Hecke algebras and characters of parabolic type of finite groups with (B,N)-pairs. Publ. Math., Inst. Hautes Etud. Sci. **40** (1972), 81-116.

[16] Curtis, C.W., and Reiner, I.: Methods of Representation Theory with Applications to Finite Groups and Orders I. J. Wiley & Sons, New York 1981.

[17] Curtis, C.W., and Reiner, I.: Methods of Representation Theory with Applications to Finite Groups and Orders II. J. Wiley & Sons, New York 1987.

[18] Freitag, E.: Siegelsche Modulfunktionen. Grundlehren Math. Wiss. **254**, Springer-Verlag, Berlin-Heidelberg-New York 1983.

[19] Gunning, R.C.: Lectures on Modular Forms. Princeton University Press, Princeton, N.J. 1962.

[20] Hecke, E.: Über Modulfunktionen und die Dirichletschen Reihen mit Eulerscher Produktentwicklung I, II. Math. Ann. **114** (1937), 1-28, 316-351; Math. Werke, 644-707.

[21] Herrmann, O.: Über Hilbertsche Modulfunktionen und die Dirichletschen Reihen mit Eulerscher Produktentwicklung. Math. Ann. **127** (1954), 357-400.

[22] Hua, L.K., and Reiner, I.: On the generators of the symplectic modular group. Trans. Am. Math. Soc. **65** (1949), 415-426.

[23] Hungerford, T.W.: Algebra. Springer-Verlag, New York-Heidelberg-Berlin 1974.

[24] Iwahori, N.: On the structure of the Hecke ring of a Chevalley group over a finite field. J. Fac. Sci., Univ. Tokyo **10** (1964), 215-236.

[25] Iwahori, N., and Matsumoto, H.: On some Bruhat decomposition and the structure of the Hecke rings of p-adic Chevalley groups. Publ. Math., Inst. Hautes Etud. Sci. **25** (1965), 5-48.

[26] Jacobson, N.: Basic Algebra I. Freeman, San Francisco 1974.

[27] Kazhdan, D., and Lusztig, G.: Representations of Coxeter groups and Hecke algebras. Invent. Math. **53** (1979), 165-184.

[28] Kazhdan, D., and Lusztig, G.: Proof of the Deligne-Langlands conjecture for Hecke algebras. Invent. Math. **87** (1987), 153-215.

[29] Kleine Schonefeld, T.: Hecke-Algebren der Dieder-Gruppe. Diplomarbeit, Münster 1984.

[30] Klingen, H.: Über die Erzeugenden gewisser Modulgruppen. Nachr. Akad. Wiss. Göttingen 1956, 173-185.

[31] Koecher, M.: Zur Theorie der Modulformen n-ten Grades I. Math. Z. **59** (1954), 399-416.

[32] Koecher, M.: Matrices over $\mathbb{Z}$. Manuscript, Münster 1985.

[33] Koecher, M.: Elliptische Funktionen und Modulformen. Lecture Notes, Münster 1989.

[34] Krieg, A.: Modular Forms on Half-Spaces of Quaternions. Lect. Notes Math. **1143**, Springer-Verlag, Berlin-Heidelberg-New York 1985.

[35] Krieg, A.: Das Vertauschungsgesetz zwischen Hecke-Operatoren und dem Siegelschen Φ-Operator. Arch. Math. **46** (1986), 323-329.

[36] Krieg, A.: Maschke's Theorem for Hecke-Algebras. J. Indian Math. Soc., New Ser. **50** (1986), 139-147.

[37] Krieg, A.: The elementary divisor theory over the Hurwitz order of integral quaternions. Linear Multilinear Algebra **21** (1987), 325-344.

[38] Krieg, A.: The Hecke-Algebras related to the Unimodular and Modular Group over the Hurwitz Order of Integral Quaternions. Proc. Indian Acad. Sci., Math. Sci. **97** (1987), 201-227.

[39] Krieg, A.: Eisenstein-Series on Real, Complex, and Quaternionic Half-Spaces. Pac. J. Math. **133** (1988), 315-354.

[40] Krieg, A.: Associative Triple Systems of Hecke-Type. Algebras Groups Geom. **5** (1988), 341-357.

[41] Kronecker, L: Über bilineare Formen. J. Reine Angew. Math. **68** (1868), 273-285; Werke I, 143-162.

[42] Lang, S.: Algebra. Addison Wesley, Reading 1965.

[43] Maaß, H.: Die Primzahlen in der Theorie der Siegelschen Modulfunktionen. Math. Ann. **124** (1951), 87-122.

[44] Maaß, H.: Lectures on Modular Forms of One Complex Variable. Tata Inst. Fund. Res., Bombay 1964; reprinted, Springer-Verlag, Berlin-Heidelberg-New York 1983.

[45] Maaß, H.: Siegel's Modular Forms and Dirichlet Series. Lect. Notes Math. **216**, Springer-Verlag, Berlin-Heidelberg-New York 1971.

[46] Matsumoto, H.: Générateurs et rélations des groupes de Weyl generalisés. C. R. Acad. Sci. Paris **258** (1964), 3419-3422.

[47] Miyake, T.: Modular Forms. Springer-Verlag, Berlin-Heidelberg-New York 1989.

[48] Mordell, L.J.: On Mr. Ramanujan's empirical expansions of modular functions. Proc. Camb. Phil. Soc. **19** (1920), 117-124.

[49] Newman, M.: Integral Matrices. Academic Press, New York-London 1972.

[50] Ogg, A.: Modular Forms and Dirichlet Series. Benjamin, New York 1969.

[51] Passman, D.S.: The Algebraic Structure of Group Rings. J. Wiley & Sons, New York 1977.

[52] Petersson, H.: Konstruktion der sämtlichen Lösungen einer Riemannschen Funktionalgleichung durch Dirichlet-Reihen mit Eulerscher Produktentwicklung I, II, III. Math. Ann. **116** (1939), 401-412; Math. Ann. **117** (1940), 39-64; Math. Ann. **117** (1940), 277-300.

[53] Satake, I.: Theory of spherical functions on reductive algebraic groups over p-adic fields. Publ. Math., Inst. Hautes Etud. Sci. **18** (1963), 5-69.

[54] Scharlau, W.: Quadratic and Hermitian Forms. Grundl. Math. Wiss. **270**, Springer-Verlag, Berlin-Heidelberg-New York 1985.

[55] Serre, J.P.: A Course in Arithmetic. Springer-Verlag, Berlin-Heidelberg-New York 1973.

[56] Shimura, G.: Sur les integrales attachées aux formes automorphes. J. Math. Soc. Japan **11** (1959), 291-311.

[57] Shimura, G.: On modular correspondences for $Sp(N,\mathbb{Z})$ and their congruence relations. Proc. Natl. Acad. Sci. USA **49** (1963), 824-828.

[58] Shimura, G.: Introduction to the Arithmetic Theory of Automorphic Functions. Iwanami Publishers and Princeton University Press, Tokyo-Princeton, N.J. 1971.

[59] Sugawara, M.: On the Transformation Theory of Siegel's Modular Group of the n-th. Degree. Proc. imp. Acad. Japan **13** (1937), 335-338.

[60] Sugawara, M.: An Invariant Property of Siegel's Modular Function. Proc. imp. Acad. Japan **14** (1938), 1-3.

[61] Sylvester, J.J.: Théorème sur les déterminants. Nouv. Ann. Math. **13** (1854), 305; Coll. Math. Papers II, 28.

[62] Tamagawa, T.: On the ζ-functions of a division algebra. Ann. Math., II. Ser. **77** (1963), 387-405.

[63] Terras, A.: Harmonic Analysis on Symmetric Spaces an Applications I. Springer-Verlag, New York-Berlin-Heidelberg 1985.

[64] Terras, A.: Harmonic Analysis on Symmetric Spaces and Applications II. Springer-Verlag, New York-Berlin-Heidelberg 1988.

[65] Witt, E.: Eine Identität zwischen Modulformen zweiten Grades. Abh. Math. Sem. Hansische Univ. **14** (1941), 323-337.

[66] Zharkovskaya, N.A.: The Siegel operator and Hecke operators. Functional Analyis Appl. **8** (1974), 113-120.

Aloys Krieg
Mathematisches Institut der
Westfälischen Wilhelms-Universität
Einsteinstr. 62
D-4400 Münster
Federal Republic of Germany

General instructions to authors for

PREPARING REPRODUCTION COPY FOR MEMOIRS

For more detailed instructions send for AMS booklet, "A Guide for Authors of Memoirs." Write to Editorial Offices, American Mathematical Society, P.O. Box 6248, Providence, R.I. 02940.

MEMOIRS are printed by photo-offset from camera copy fully prepared by the author. This means that, except for a reduction in size of 20 to 30%, the finished book will look exactly like the copy submitted. Thus the author will want to use a good quality typewriter with a new, medium-inked black ribbon, and submit clean copy on the appropriate model paper.

Model Paper, provided at no cost by the AMS, is paper marked with blue lines that confine the copy to the appropriate size. Author should specify, when ordering, whether typewriter to be used has **PICA**-size (10 characters to the inch) or **ELITE**-size type (12 characters to the inch).

Line Spacing — For best appearance, and economy, a typewriter equipped with a half-space ratchet — 12 notches to the inch — should be used. (This may be purchased and attached at small cost.) Three notches make the desired spacing, which is equivalent to 1-1/2 ordinary single spaces. Where copy has a great many subscripts and superscripts, however, double spacing should be used.

Special Characters may be filled in carefully freehand, using dense black ink, or **INSTANT** ("rub-on") **LETTERING** may be used. AMS has a sheet of several hundred most-used symbols and letters which may be purchased for $5.

Diagrams may be drawn in black ink either directly on the model sheet, or on a separate sheet and pasted with rubber cement into spaces left for them in the text. Ballpoint pen is not acceptable.

Page Headings (Running Heads) should be centered, in CAPITAL LETTERS (preferably), at the top of the page — just above the blue line and touching it.

LEFT-hand, EVEN-numbered pages should be headed with the AUTHOR'S NAME;

RIGHT-hand, ODD-numbered pages should be headed with the TITLE of the paper (in shortened form if necessary).

Exceptions: PAGE 1 and any other page that carries a display title require NO RUNNING HEADS.

Page Numbers should be at the top of the page, on the same line with the running heads.

LEFT-hand, EVEN numbers — flush with left margin;

RIGHT-hand, ODD numbers — flush with right margin.

Exceptions: PAGE 1 and any other page that carries a display title should have page number, centered below the text, on blue line provided.

FRONT MATTER PAGES should be numbered with Roman numerals (lower case), positioned below text in same manner as described above.

MEMOIRS FORMAT

It is suggested that the material be arranged in pages as indicated below.
Note: Starred items (*) are requirements of publication.

Front Matter (first pages in book, preceding main body of text).

Page i — *Title, *Author's name.

Page iii — Table of contents.

Page iv — *Abstract (at least 1 sentence and at most 300 words).

Key words and phrases, if desired. (A list which covers the content of the paper adequately enough to be useful for an information retrieval system.)

**1980 Mathematics Subject Classification (1985 Revision)*. This classification represents the primary and secondary subjects of the paper, and the scheme can be found in Annual Subject Indexes of MATHEMATICAL REVIEWS beginnning in 1984.

Page 1 — Preface, introduction, or any other matter not belonging in body of text.

Footnotes: *Received by the editor date.
Support information — grants, credits, etc.

First Page Following Introduction - Chapter Title (dropped 1 inch from top line, and centered). Beginning of Text.

Last Page (at bottom) - Author's affiliation.